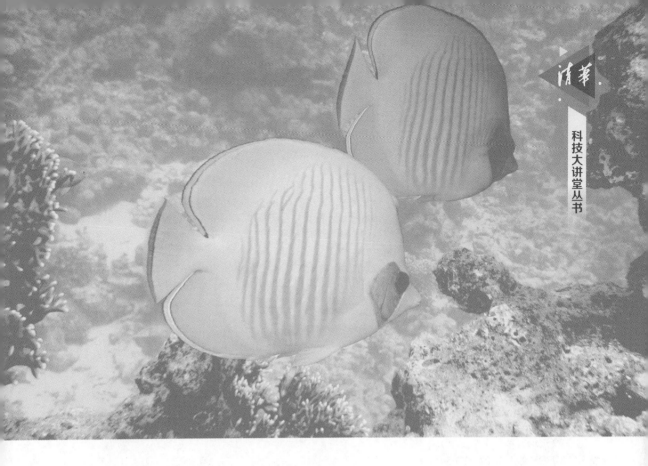

清華

科技大讲堂丛书

Java EE 轻量级框架整合开发

——Spring+Spring MVC+MyBatis 微课版

彭之军 ◎ 主编
陈立为 刘波 ◎ 副主编

清華大学出版社
北京

内 容 简 介

本书理论结合实践，全面、系统地介绍了 Spring、Spring MVC、MyBatis 框架开发后端应用程序的知识，通过简单的网上书店案例详细地介绍了项目开发的一般过程和方法，以循序渐进的方式讲解了 SSM 框架的各种特性，并通过员工管理系统实例整合 Vue 前端技术和 SSM 后端开发技术，让读者能够快速掌握并学以致用。

全书共 11 章。第 1 章主要内容为 Java EE 开发简介和网上书店案例；第 2~8 章主要讲解 Spring 框架、Spring MVC 和 MyBatis 的相关知识；第 9 章主要内容为 Spring 框架的事务管理；第 10 章主要讲解 Vue 前端框架开发；第 11 章主要内容为 Element+SSM 框架项目开发。本书偏重于实践教学，在讲解理论知识的同时，通过一些典型案例让读者了解理论知识在实际环境中的应用，并对易混淆和较难理解的知识点做了重点分析，以加深读者对知识的理解。

本书附有教学视频、源代码、课件、教学大纲等配套资源，可以作为大学计算机及相关专业的教材和教学参考用书，也可以作为 Java 技术初学者的培训教材，同时适用于广大 Java EE 应用开发人员进行查阅与使用。

图书在版编目（CIP）数据

Java EE轻量级框架整合开发：Spring+Spring MVC+MyBatis：微课版/彭之军主编. —北京：清华大学出版社，2023.7（2024.7重印）
（清华科技大讲堂丛书）
ISBN 978-7-302-62765-4

Ⅰ. ①J… Ⅱ. ①彭… Ⅲ. ①JAVA语言－程序设计 Ⅳ. ①TP312.8

中国国家版本馆CIP数据核字(2023)第031794号

策划编辑：魏江江
责任编辑：王冰飞
封面设计：刘 键
责任校对：时翠兰
责任印制：杨 艳

出版发行：清华大学出版社
网 址：https://www.tup.com.cn，https://www.wqxuetang.com
地 址：北京清华大学学研大厦 A 座 邮 编：100084
社 总 机：010-83470000 邮 购：010-62786544
投稿与读者服务：010-62776969，c-service@tup.tsinghua.edu.cn
质 量 反 馈：010-62772015，zhiliang@tup.tsinghua.edu.cn
课 件 下 载：https://www.tup.com.cn，010-83470236
印 装 者：小森印刷霸州有限公司
经 销：全国新华书店
开 本：185mm×260mm 印 张：19.75 字 数：481 千字
版 次：2023 年 7 月第 1 版 印 次：2024 年 7 月第 3 次印刷
印 数：3001～5000
定 价：59.80 元

产品编号：080902-01

前言 Preface

党的二十大报告中指出：教育、科技、人才是全面建设社会主义现代化国家的基础性、战略性支撑。必须坚持科技是第一生产力、人才是第一资源、创新是第一动力，深入实施科教兴国战略、人才强国战略、创新驱动发展战略，这三大战略共同服务于创新型国家的建设。高等教育与经济社会发展紧密相连，对促进就业创业、助力经济社会发展、增进人民福祉具有重要意义。

随着大数据和人工智能技术的兴起，Java 后端开发技术似乎很少被关注，但这并不意味着 Java 后端开发技术不再重要。目前，Spring 框架是以 Java 语言为主的后端开发技术作为核心技术，加上从 Spring 框架衍生而来的 Spring MVC、Spring Boot、Spring Cloud 等框架，已经成为互联网中最重要的软件基础设施之一。

本书面向 Java 后端开发工程师和有志于成为前后端都精通的全栈开发工程师。本书系统而全面地介绍了 Spring 框架、Spring MVC 框架、MyBatis 框架、Vue 前端开发和基于 Vue 的 Element 前端框架等内容，由浅入深、循序渐进，结合大量的案例进行讲解，有利于加深读者对知识的理解。读者通过本书可以掌握 SSM 框架的相关知识，为后端开发打下坚实的基础，同时本书的 Vue 开发部分和 Element 框架也有助于前端开发能力的提高。

全书共 11 章。第 1 章主要内容为 Java EE 开发简介和网上书店案例；第 2~4 章主要讲解 Spring 框架的相关知识，内容包括 Spring IoC、Spring AOP 和 Spring JDBC；第 5、6 章主要讲解 Spring MVC 的相关知识，内容包括 Spring MVC 的工作原理、体系结构、注解详解和重构案例；第 7、8 章主要讲解 MyBatis 的相关知识，内容包括 MyBatis 的框架基础、MyBatis 开发流程和动态 SQL；第 9 章主要内容为 Spring 框架的事务管理；第 10 章主要讲解 Vue 前端框架开发，内容包括 Vue 入门、Vue 常用指令、绑定属性及事件、渲染和异步操作；第 11 章主要内容为 Element+SSM 框架项目开发。

本书特色如下：

（1）内容全面

本书将理论与实践相结合，全面介绍了 Spring 框架、Spring MVC 框架和 MyBatis 框架三大框架（SSM 框架），以及 Vue 前端开发和基于 Vue 的 Element 前端框架等内容，较难理解的知识点都配合图示和代码实例进行了详细讲解。

（2）循序渐进

本书从最基本的 Java Web 开发网上书店案例开始入手，逐步深入 Spring 框架和 Spring MVC 框架，最后切换到使用 MyBatis 访问数据库。这样可以让读者快速地掌握 SSM 的必需知识，并能在实际开发中加以使用。

（3）实例丰富

本书偏重于实践教学，对于理论知识点，书中都有代码实例予以讲解。

（4）经验传授

本书是基于编著者多年的开发和教学经验凝练而成。本书主编彭之军老师曾在两家具有 CMMI5 级以上的公司任 Java 高级软件开发工程师，后在 IT 培训机构和高校计算机系工作，目前为广东邮电职业技术学院移动通信学院副教授，具有十年以上的教学经验。陈立为老师从事 IT 职业培训达十五年以上。刘波老师是传智播客广州分公司的 Java 高级讲师，二十年来培训的学员达上千人。三位作者培训的学员遍布珠三角大大小小的 IT 公司。

本书彭之军编写了第 1、5、6 章，陈立为编写了第 4、7、8、9 章，刘波编写了第 2、3、10、11 章。全书由彭之军统稿。

为便于教学，本书提供丰富的配套资源，包括教学大纲、教学课件、电子教案、程序源码和微课视频。

资源下载提示

课件等资源：扫描封底的“课件下载”二维码，在公众号“书圈”下载。
数据文件等资源：扫描目录上方的二维码下载。
视频等资源：扫描封底的文泉云盘防盗码，再扫描书中相应章节的二维码，可以在线学习。

由于编者水平有限，书中难免会有不足之处，敬请广大读者批评指正。

编者

2023 年 7 月

目 录 Contents

资源下载

第 1 章　初探 Java EE 开发　/1

1.1　Spring 框架的发展　/2
1.2　开发环境的准备　/2
1.2.1　JDK 的安装与配置　/3
1.2.2　Tomcat 的安装与配置　/4
1.3　网上书店项目案例　/7
1.3.1　GoodBook 网上书店需求　/7
1.3.2　GoodBook 网上书店实体关系图　/7
1.3.3　GoodBook 网上书店三层架构　/9
1.3.4　三层架构代码实现　/9
1.3.5　JUnit 测试框架　/14
1.3.6　三层架构业务逻辑层实现　/19
1.3.7　三层架构 Web 层实现　/21
1.4　本章小结　/25
习题 1　/25

第 2 章　Spring IoC　/26

2.1　Spring 框架简介　/27
2.2　Spring 的体系结构　/29
2.2.1　Spring 环境简介　/29
2.2.2　IoC 容器入门　/30
2.2.3　Bean 标签的配置　/33
2.2.4　scope 属性值　/34

2.3 依赖注入 /36
2.3.1 依赖注入简介 /36
2.3.2 构造器注入 /36
2.3.3 使用 set 注入 /38
2.3.4 使用 p 命名空间 /40
2.4 基于注解方式的 IoC /41
2.4.1 使用注解 /41
2.4.2 扫描基包 /41
2.4.3 IoC 容器中的注解 /43
2.5 依赖关系的注解 /44
2.5.1 按类型匹配注入 /44
2.5.2 按名字匹配注入 /45
2.5.3 注入简单类型 /46
2.6 本章小结 /48
习题 2 /48

第 3 章 Spring AOP /49

3.1 Spring AOP 概述 /50
3.1.1 AOP 的概念 /50
3.1.2 AOP 中类与切面的关系 /51
3.1.3 AOP 的应用场景 /51
3.2 动态代理模式 /52
3.2.1 代理模式对象 /52
3.2.2 JDK 动态代理 /53
3.3 AOP 的实现 /55
3.3.1 AOP 的常用增强类型 /55
3.3.2 AspectJ 表达式 /56
3.3.3 使用 XML 配置方式实现 AOP /58
3.3.4 使用注解方式实现 AOP /62
3.4 本章小结 /66
习题 3 /67

第 4 章 Spring JDBC /68

4.1 Spring JDBC 简介 /69
4.2 JdbcTemplate 各种方法的使用 /69

4.2.1 execute 方法 /69
4.2.2 update 方法 /72
4.2.3 query 方法 /75
4.3 数据源的配置 /79
4.3.1 DBCP 数据源 BasicDataSource 的使用 /80
4.3.2 C3P0 数据源 ComboPooledDataSource 的使用 /81
4.3.3 使用属性文件读取数据库连接信息 /82
4.4 本章小结 /84
习题 4 /84

第 5 章 Spring MVC /85

5.1 Spring MVC 简介 /86
5.2 第一个 Spring MVC 案例 /86
5.3 Spring MVC 的工作原理与体系结构 /90
5.3.1 Spring MVC 程序运行原理 /90
5.3.2 视图解析器 /90
5.3.3 Spring MVC 的体系结构 /92
5.4 基于注解的控制器配置 /93
5.5 Spring MVC 注解详解 /95
5.5.1 在类前注解 /95
5.5.2 RequestMapping 注解属性 /95
5.5.3 cURL 工具软件 /97
5.6 本章小结 /99
习题 5 /99

第 6 章 基于 Spring MVC 的网上书店重构 /100

6.1 会员模块实现 /101
6.1.1 用户信息显示功能 /101
6.1.2 会员注册和登录功能 /104
6.2 图书模块实现 /114
6.3 购物车模块实现 /120
6.4 订单模块实现 /125
6.5 本章小结 /133
习题 6 /133

第 7 章　MyBatis 框架入门　/134

7.1　MyBatis 框架简介　/135

7.2　MyBatis 开发环境　/136

7.2.1　MyBatis 的下载　/136

7.2.2　搭建 MyBatis 开发环境　/137

7.3　MyBatis 开发流程　/137

7.3.1　MyBatis 基本开发流程　/137

7.3.2　第一个 MyBatis 项目　/138

7.3.3　MyBatis 工作流程　/142

7.4　使用 MyBatis 实现增、删、改、查操作　/143

7.4.1　使用 selectOne 方法查询单个员工　/143

7.4.2　使用 insert 方法添加员工　/145

7.4.3　使用 delete 方法删除员工　/147

7.4.4　使用 update 方法修改员工　/149

7.4.5　使用工具类 MyBatisUtil 减少冗余　/153

7.5　parameterType 输入参数　/156

7.6　ResultMap 结果映射　/159

7.7　接口动态代理　/161

习题 7　/164

上机练习 1　/164

第 8 章　MyBatis 框架深入　/165

8.1　动态查询　/166

8.1.1　<if>标签　/166

8.1.2　<where>标签　/170

8.1.3　<choose>标签　/171

8.1.4　<foreach>标签　/174

8.1.5　<sql>标签　/175

8.2　多表之间的关系　/176

8.3　一对多查询　/176

8.4　多对一查询　/182

8.5　自连接查询　/187

8.5.1　以多对一的方式实现自连接　/187

8.5.2　以一对多的方式实现自连接　/190

8.6 多对多查询 /193
8.7 分页查询 /200
8.7.1 MyBatis 分页查询原理 /200
8.7.2 使用 PageHelper 实现分页 /201
8.7.3 分页实践 /202
8.8 缓存 /205
8.8.1 一级缓存 /206
8.8.2 二级缓存 /206
习题 8 /207
上机练习 2 /208

第 9 章 Spring 事务管理 /209

9.1 事务管理的概念 /210
9.2 Spring 事务管理的核心接口 /210
9.2.1 TransactionDefinition 接口 /211
9.2.2 TransactionStatus 接口 /216
9.2.3 PlatformTransactionManager 接口 /217
9.3 声明式事务 /219
9.3.1 编程式和声明式事务的区别 /219
9.3.2 基于 XML 配置文件的事务管理 /219
9.3.3 注解式事务管理 /224
习题 9 /226
上机练习 3 /227

第 10 章 前端框架 Vue 基础 /228

10.1 Vue 简介 /229
10.2 IntelliJ IDEA 开发环境 /230
10.3 Vue 快速入门 /233
10.4 Vue 常用指令 /236
10.5 绑定属性 /238
10.6 绑定事件 /239
10.7 条件渲染 /241
10.8 循环渲染 /243
10.9 双向绑定 /245

10.10 Vue 的 AJAX 异步操作 /248
10.11 本章小结 /253
习题 10 /253

第 11 章 Element+SSM 开发员工管理模块 /255

11.1 Maven 基础 /256
11.1.1 为什么要学习 Maven /256
11.1.2 Maven 基本概念 /256
11.1.3 Maven 的安装与配置 /257
11.1.4 在 IDEA 中配置 Maven /260
11.2 使用 Maven 搭建 SSM 环境 /261
11.2.1 创建 Maven 工程 /261
11.2.2 完善工程的目录结构 /262
11.2.3 搭建 SSM 开发环境 /263
11.2.4 在 Tomcat 中部署运行 /270
11.3 员工管理系统的实现 /271
11.3.1 项目需求 /271
11.3.2 运行效果 /271
11.3.3 数据库设计 /272
11.3.4 Lombok 插件 /274
11.3.5 实体类对象 /275
11.3.6 数据访问层 /278
11.3.7 业务层 /281
11.3.8 测试业务层 /284
11.3.9 控制器层 /285
11.4 基于 Element 框架的系统开发 /288
11.4.1 什么是 Element /288
11.4.2 Element 快速入门 /288
11.4.3 Element 第一个案例 /289
11.4.4 使用 Element 实现员工系统的表示层 /291
11.5 本章小结 /304
习题 11 /304

参考文献 /305

第 1 章 初探Java EE开发

本章学习内容

- Spring 框架的发展；
- 开发环境的准备；
- 网上书店项目案例。

1.1 Spring 框架的发展

随着软件开发技术的进步和用户需求的不断变化，Java 语言自身也在不断地成长和进化，JDK 从最初的 1.0 版本更新到现在的 18 版本。即使是 Oracle 官方设立的 Java EE 标准，如今也不再具有绝对的权威。为什么官方的标准失去了统治力？因为 Java 世界有了无数优秀的第三方开源软件，有些软件的适用性甚至超过了官方的标准架构，其中尤以 Spring 框架为翘楚。

Sun 公司于 1997 年发布的 EJB（Enterprise Java Bean）用大型分布式软件架构的思维来设计 Java 项目。大部分软件系统都是中小型的软件系统，使用 EJB 架构就像用航空母舰来解决普通的乘客运输问题：它的确很稳健，但是大多数时候是不适用的。所以 EJB 被称为重量级框架，而后出现的 Spring、Struts、Struts2、Hibernate、MyBatis 等被称为轻量级框架。

Spring 框架具有一套完整的 Java 技术生态链，已经成为事实上的 Java EE 标准。Spring 从创建之初就是为了解决企业应用程序开发中日益复杂的问题，它的出现将 Java EE 开发者从 EJB 的泥潭中解救了出来。

Spring 框架来源于企业开发实践。具有多年 C++和 J2EE 开发经验的 Rod Johnson 在《Expert One-on-One J2EE 设计和开发》中分享了 Spring 框架的设计原型。凭借着丰富的企业实践功底，他与团队所设计开发的 Spring 框架被广大 Java EE 开发者所追捧，创造了 Java 语言的春天。

在大中型软件项目中，软件架构所奉行的一个重要的原则是“高内聚、低耦合”。Spring 框架的主要优势之一是其优秀的分层架构设计，降低了模块之间的耦合度。分层架构允许开发者选择使用哪一个组件，同时为 Java EE 应用程序开发提供集成的框架。

视频讲解

1.2 开发环境的准备

计算机软件是一门实践性很强的学科。就像学习游泳一样，只是站在岸上看别人游泳永远也无法学会游泳。学习编程就必须要亲自动手编写代码，解决一个又一个错误，编写代码达到一定数量，然后才能由量变到质变，成为一名合格的程序员。除了天才，没有人能仅仅通过看书就能掌握软件编程的奥秘。

工欲善其事，必先利其器。接下来安装必要的开发工具和软件。

需要安装的软件列表如表 1-1 所示。

表 1-1　需要安装的软件列表

序号	软件名称	建议版本	作　用	备　注
1	JDK	JDK8	Java 开发工具	必需
2	Tomcat	8.0 或 8.5	Web 服务器	必需
3	MySQL	8.0 以上	数据库服务器	必需
4	MyEclipse	2014	集成开发工具	非必需，4、5、6 三选一
5	IntelliJ IDEA	2019.12	集成开发工具	非必需，4、5、6 三选一
6	Eclipse EE	2021.12	集成开发工具	非必需，4、5、6 三选一
7	Maven	3.6.3	项目构建工具	非必需

1.2.1　JDK 的安装与配置

在 Java 或 Java EE 的开发中，JDK 是基础软件，为此需要首先安装 JDK。

JDK 的安装步骤如下所示。

步骤 1：下载并安装 JDK。下载网址为 http://www.oracle.com/technetwork/java/javase/downloads/index.html，但是该地址只有最新的 JDK 版本。目前官方发布的最新版本是 JDK18，与其配套的很多软件并没有跟上它的脚步，所以不建议在企业开发中立即使用，至少等一两年成熟后才可以被企业所采用。已发布的所有 JDK 旧版本的下载网址为 http://www.oracle.com/technetwork/java/javase/archive-139210.html。目前 JDK8 是企业用得比较多的版本，因为它兼容性好、稳定性高。

本书要求安装 JDK8，注意版本和平台的选择，在此介绍 Windows 7 或 Windows 10 平台下 JDK 的安装。JDK8 的下载地址为 https://www.oracle.com/java/technologies/javase/javase8u211-later-archive-downloads.html。选择 Windows 平台下 64 位的版本，如图 1-1 所示。

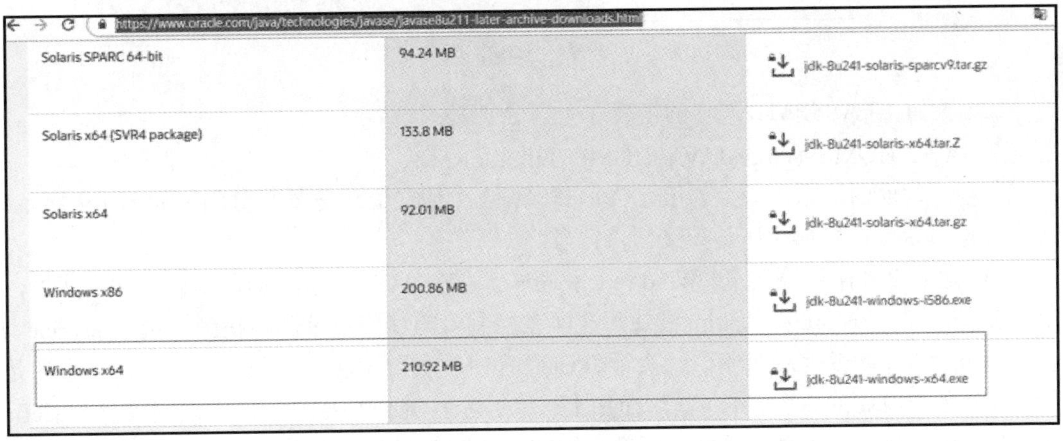

https://www.oracle.com/java/technologies/javase/javase8u211-later-archive-downloads.html

Solaris SPARC 64-bit	94.24 MB	jdk-8u241-solaris-sparcv9.tar.gz
Solaris x64 (SVR4 package)	133.8 MB	jdk-8u241-solaris-x64.tar.Z
Solaris x64	92.01 MB	jdk-8u241-solaris-x64.tar.gz
Windows x86	200.86 MB	jdk-8u241-windows-i586.exe
Windows x64	210.92 MB	jdk-8u241-windows-x64.exe

图 1-1　JDK 下载

下载完成后，按照标准安装方式，每个页面保持默认设置，一直单击“下一步”按钮就可以完成安装了。

安装的默认路径为 C:\Program Files\Java\jdk1.8.0_241，注意要确认这个地址是否正确，如果安装时自定义文件夹，需要改成自定义的文件夹目录。

步骤 2：环境变量的配置。这一步尤其重要，很多初学者和经验不多的开发者都会犯错，导致后面的开发中出现各种各样的问题。Java 在 Windows 平台下用到的环境变量主要有 3 个，即 JAVA_HOME、CLASSPATH 和 PATH。

（1）新建 JAVA_HOME 环境变量，指向 JDK 的安装路径，如图 1-2 所示。

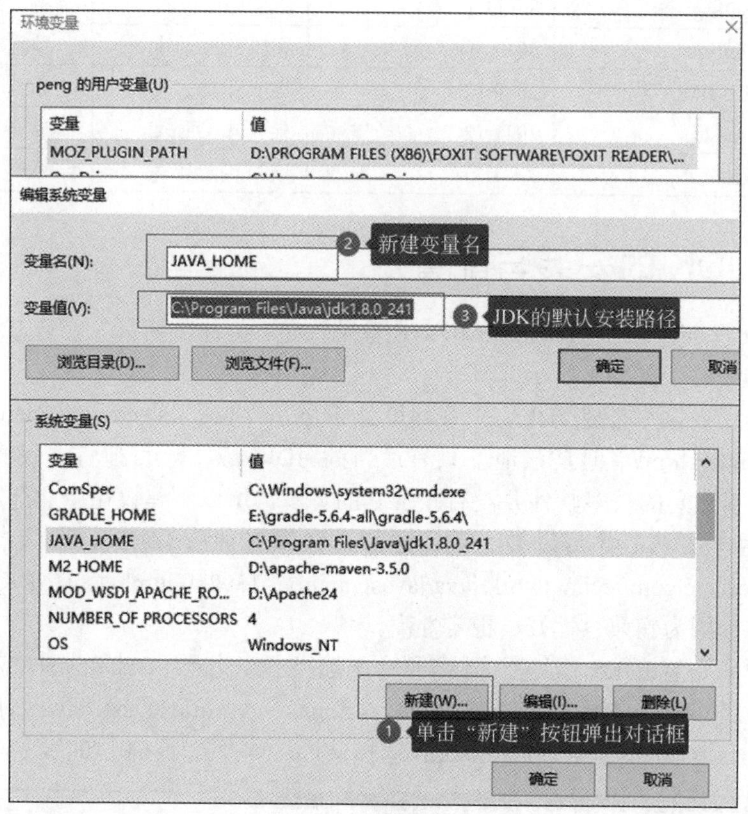

图 1-2 JAVA_HOME 环境变量

（2）配置 CLASSPATH。它的值有 3 个，分别是：

.;%JAVA_HOME%\lib;%JAVA_HOME%\lib\tools.jar

注意，最前面的第一个点号表示当前目录，3 个值之间用分号来分隔，如图 1-3 所示。这里所有的标点符号都应该是英文标点符号。

（3）建立 PATH 环境变量。Windows 系统本就存在 PATH 变量，但需要修改这个变量，使 Windows 能知道 JDK 的 bin 目录。这样在控制台的任何目录都能找到并执行 javac.exe 和 java.exe 文件，否则后面编译、执行 Java 程序时就需要再输入一长串路径。设置方法是保留原来的 PATH 的内容，并在其后面加上 %JAVA_HOME%\bin。然后可以编写一个 Hello world 的 Java 程序，手动编译并运行它，如果它能正确运行，则表示 JDK 环境变量配置正确完成。

1.2.2 Tomcat 的安装与配置

接下来安装并配置 Web 服务器 Tomcat。

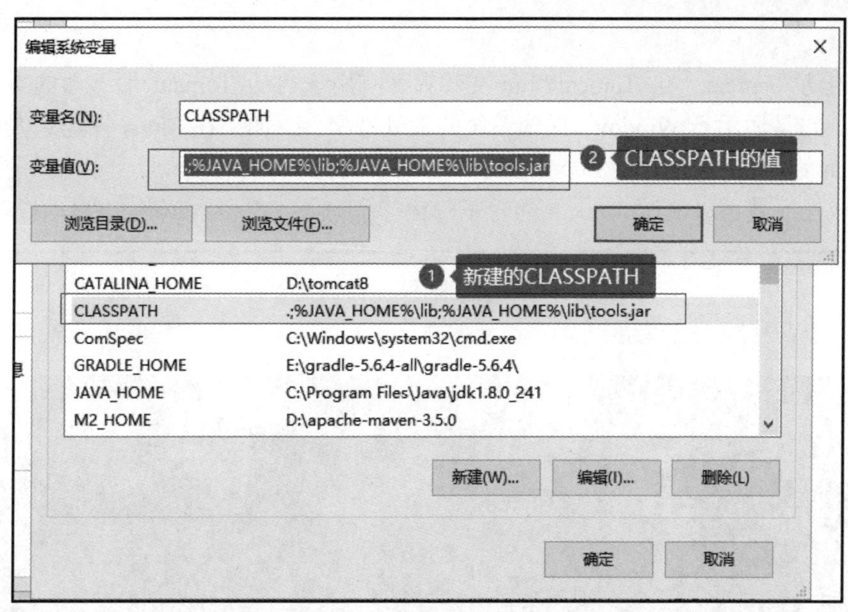

图 1-3 CLASSPATH 环境变量

（1）下载 Tomcat。下载地址为 https://tomcat.apache.org/index.html。根据计算机的系统下载对应的版本，在此下载的是 Windows 64 位的 zip 包。现在最新版本是 Tomcat 10。但和 JDK 的选择一样，降级选择一个版本，Tomcat 8 版本是较为稳定的版本。图 1-4 为 Tomcat 下载页面。

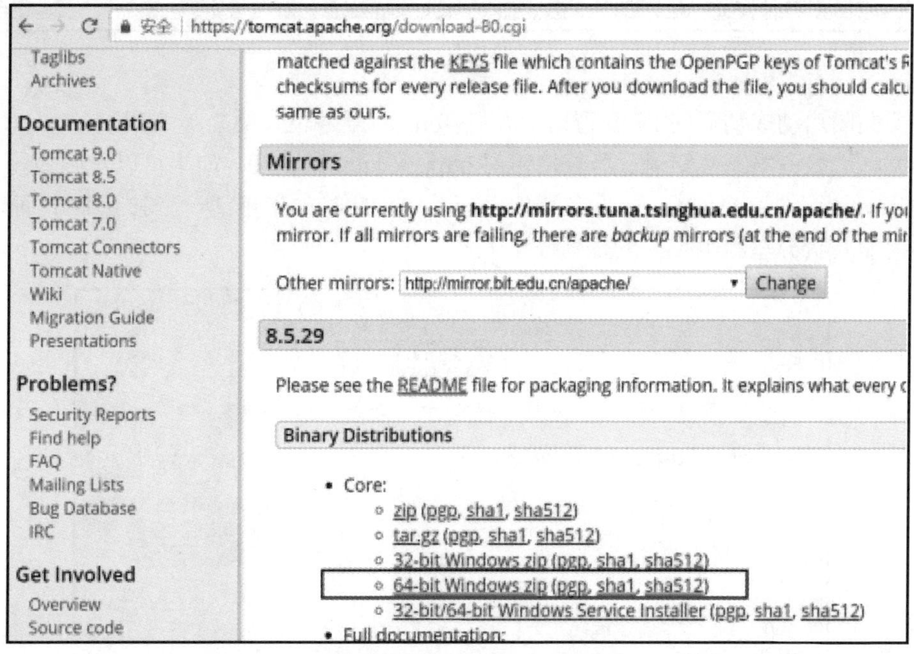

图 1-4 Tomcat 下载页面

（2）配置 Tomcat。如果 JDK 配置正确，Tomcat 几乎不用额外配置，解压后即可使用。建议解压缩后的文件夹放在非系统盘（如 D 盘）根目录下。本书示例放在 D:\Tomcat8

目录下。

（3）启动 Tomcat。在 Tomcat8\bin\文件夹下有用于启动 Tomcat 服务器的文件，名为 startup.bat，它是适用于 Windows 操作系统的批处理文件，如果在 Linux 操作系统下则要运行 startup.sh 的 shell 脚本。

启动成功后就可以测试 Tomcat 能否正确运行了。图 1-5 表示 Tomcat 启动成功。

图 1-5　Tomcat 启动成功

不要关闭启动成功后的 DOS 窗口，否则 Tomcat 服务也会随之被关闭。

（4）测试服务器。打开浏览器直接访问 http://localhost:8080，如果可以看到如图 1-6 所示的 Tomcat 界面，表示 Tomcat 服务器启动成功了。此时如果申请一个独立 IP 地址，就可以将编写的网页分享给世界上的任何一个人。

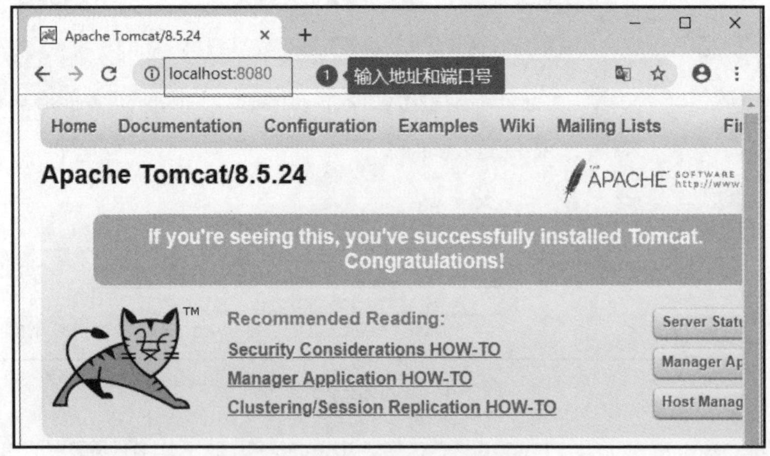

图 1-6　本地 Tomcat 服务器主页

MySQL 8.x 数据库服务器和 MyEclipse、IDEA 等开发工具的安装，读者可以在对应软件的官网上下载后根据安装向导进行安装，篇幅所限具体步骤不再赘述。

1.3 网上书店项目案例

视频讲解

一个小型软件系统的创建过程会包括如下问题：软件系统如何从 0 到 1 进行开发的？如何进行前期的设计？如何设计数据库？如何设计系统的分层架构？本节将通过一个小型的网上书店 GoodBook 案例来介绍以 Spring MVC 技术为主的 Java Web 项目。

1.3.1 GoodBook 网上书店需求

按照敏捷软件开发的原则快速地整理客户的需求，将其编写成用户故事（User Story）表，关于用户故事的详细介绍可以参见敏捷软件开发相关内容。需求列表如表 1-2 所示。

表 1-2　网上书店的用户故事

故事编号	故事内容
1	客户可以通过系统注册并登录自己的账户
2	客户可以修改自己的注册信息
3	客户可以查询销售的图书
4	客户可以将订购的图书放入购物车中并提交订单
5	客户可以查询历史订单并取消未完成订单
6	网站工作人员可以对图书进行上架和下架处理，并进行图书信息维护
7	网站工作人员可以查询和统计销售记录
8	网站工作人员可以对订单进行确认和查询
9	网站工作人员可以接受发货请求或者因缺货拒绝请求
10	系统管理员可以维护客户和网站工作人员信息
11	系统可以根据用户信息推荐图书

以上是目前初步列出的功能，后续随着不断地深入学习，可以再添加新的功能。下面根据需求列表设计网上书店。

1.3.2 GoodBook 网上书店实体关系图

首先根据 1.3.1 节的用户故事分析系统中的实体及其属性，然后根据分析画出实体关系图。经过简单分析，标识出了会员、订单、出版社和图书 4 个实体，如图 1-7 所示。

由图 1-7 的实体关系图可以将此系统初步划分为以下 4 个子模块。

模块 1：会员模块；

模块 2：图书模块；

模块 3：订单模块；

模块 4：出版社模块。

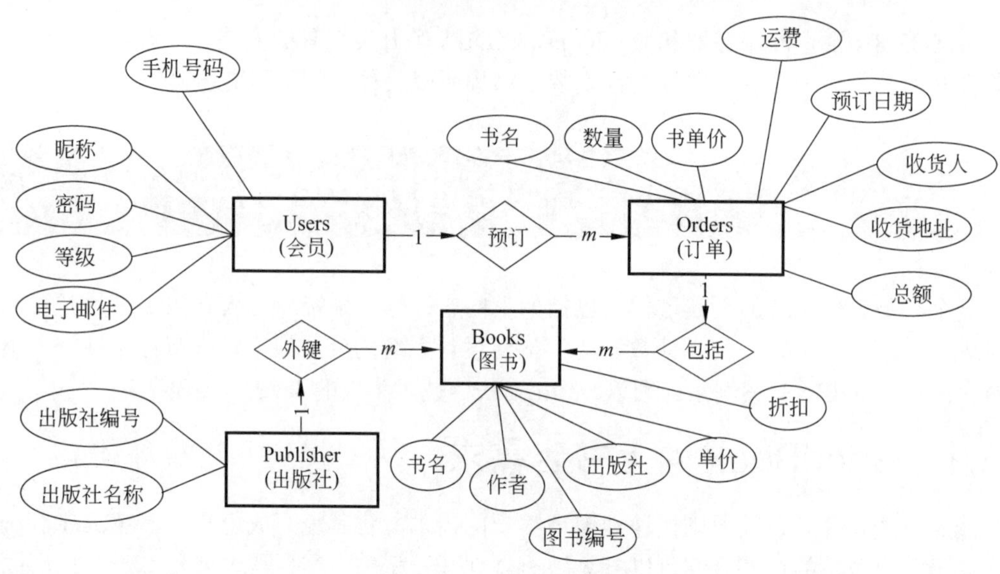

图 1-7　网上书店实体关系图

在以上 4 个模块中，图书模块和订单模块最为复杂，会员模块复杂性稍低，最简单的是出版社模块。由于图书模块比会员模块相对复杂很多，遵循由易到难的学习规律，下面从较为简单的会员模块开始介绍。会员是指网上书店的注册用户，因此也将会员称为用户。

根据网上书店实体关系图，将其从概念模型转成物理模型，也就是在 MySQL 数据库服务器中建立数据库和数据表，数据库名为 bookDb。用户表结构如表 1-3 所示。

表 1-3　用户表结构

列　名	数据类型	长　度	是否为空	是否主键
Userid	int	11	否	是
Username	varchar	20	否	否
Password	varchar	50	否	否
Email	varchar	50	否	否
Phone	varchar	30	否	否

出于简化的目的，精简了某些属性，如地址、用户等级等。

建表的 SQL 语句如下，表名为 tbuser。

```
DROP TABLE IF EXISTS 'tbuser';
CREATE TABLE tbuser (
  userid int AUTO_INCREMENT  PRIMARY key,
  username varchar(20) DEFAULT NULL,
  'password' varchar(50) DEFAULT NULL,
  Email varchar(50) DEFAULT NULL,
  Phone varchar(30) DEFAULT NULL
) ENGINE=InnoDb DEFAULT CHARSET=utf8;
```

1.3.3 GoodBook 网上书店三层架构

三层架构是 Java EE 规范中的推荐架构。传统意义上的三层架构分为表示层（UI）、业务逻辑层（BLL）和数据访问层（DAL）。在 Java EE 的开发中，三层架构具体分为表示层（Web 层）、业务逻辑层（Service）和数据访问层（DAO 层）。

三层架构是典型的架构模式（Architecture Pattern），将产品的开发细分为三层，这样做的好处是能够让每个人都发挥专长。例如，前端工程师能专注页面设计，如何吸引用户，而不用在乎业务逻辑实现；数据库工程师可以专注数据库处理，使其变得高效，而不必关注如何展示。

- Web 层：与客户端交互，包括获取用户请求、传递数据、封装数据、展示数据等。
- Service 层：进行复杂的业务处理，包括各种实际的逻辑运算。
- DAO 层：在数据访问层中主要对数据库中的表进行操作，即实现对数据表的 Select（查询）、Insert（插入）、Update（更新）、Delete（删除）等操作。如果要加入对象关系映射的元素，那么就会包括对象和数据表之间的映射，以及对象实体的持久化。

在实际开发中，有些开发者也习惯将 DAO 层再进行拆分，根据代码的不同功能划分为 4 个不同的包。

Service 层主要包括业务处理逻辑类和普通 Java 类。普通 Java 类又被称为 POJO 类，POJO（Plain Ordinary Java Object）即为简单的 Java 对象，实际就是普通的 JavaBeans，是为了避免和 EJB 混淆所创造的简称。POJO 类中有一些属性和 getter setter 方法的类，没有业务逻辑，有时可以作为 VO（value-object）或 DTO（Data Transform Object）来使用。当然，如果有一个简单的运算属性也是可以的，但一般不会有业务方法。

在本项目中建立了如图 1-8 所示的包结构。

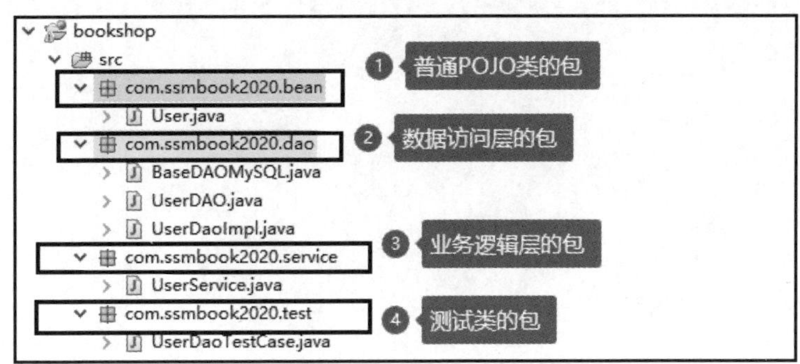

图 1-8　项目包结构图

1.3.4 三层架构代码实现

在本项目中，三层架构代码的实现步骤如下。

（1）在项目的 src 目录下建立 Java 对象。

```
package com.ssmbook2020.bean;
public class User {
    private int userid;
    private String username;
```

```
    private String passwd;
    private String email;
    private String phone;

    public User(int userid, String username, String passwd, String email,
String phone) {
        this.userid = userid;
        this.username = username;
        this.passwd = passwd;


        this.email = email;
        this.phone = phone;
    }
    public User(){
    }
    publicint getUserid() {
        return userid;
    }
    public void setUserid(int userid) {
        this.userid = userid;
    }
    public String getUsername(){
        return username;
    }
    public void setUsername(String username){
        this.username = username;
    }
    public String getPasswd(){
        return passwd;
    }
    public void setPasswd(String passwd){
        this.passwd = passwd;
    }
    public String getEmail(){
        return email;
    }
    public void setEmail(String email){
        this.email = email;
    }
public String getPhone() {
        return phone;
    }
    public void setPhone(String phone){
        this.phone = phone;
    }
}
```

（2）在数据访问层准备好通用的数据库访问类，文件名为 BaseDaoMySQL.java。

```
package com.ssmbook2020.dao;
//省略部分导包代码
/**
 * Title: 数据库操作基类,所有操作数据类都继承此类 Description:
* @author peng zj
 * @version 1.0
 */
public abstract class BaseDAOMySQL {

    static Connection conn = null;
    static PreparedStatement pstmt = null;
    static ResultSet rs = null;
    //数据库密码
    public final static String driverMYSQL = "com.mysql.jdbc.Driver";
                                                          //数据库驱动
    public final static String urlMYSQL = "jdbc:mysql://localhost:3306/
bookdb?useUnicode=true&characterEncoding=UTF-8&useSSL=false";
    public final static String dbNameMYSQL = "root";    //数据库用户名
    public final static String dbPassMYSQL = "123456";  //数据库密码

    public static ConnectiongetConnection(){
        try{
            Class.forName(driverMYSQL);
            conn = DriverManager.getConnection(urlMYSQL, dbNameMYSQL,
dbPassMYSQL);
        } catch(ClassNotFoundException e){
            e.printStackTrace();
        } catch (SQLException e){
            e.printStackTrace();
        }
        //获得数据库连接
        return conn; //返回连接
    }

    public intexeUpdate(String sql) throws SQLException{
        Connection conn = null;
        PreparedStatement pstmt = null;
        int row = 0;
        try{
            conn = getConnection();
            pstmt = conn.prepareStatement(sql);
            row = pstmt.executeUpdate();
        } catch (Exception ex){
            ex.printStackTrace();
```

```
    } finally{
        //关闭当前的连接对象
        DbUtils.closeQuietly(conn, pstmt, null);
    }
    return row;
}

/**
 * 执行insert ，update ,delete的SQL语句，数组为SQL语句参数
 *
 * @param preparedSql
 * @param param
 * @return
 */
public intexecuteSQL(String preparedSql, String[] param){
    Connection conn = null;
    PreparedStatement pstmt = null;
    int num = 0;

    /*处理SQL,执行SQL */
    try{
        conn = getConnection();                           //得到数据库连接
        pstmt = conn.prepareStatement(preparedSql); //得到PreparedStatement
        if (param != null){
            for (inti = 0; i < param.length; i++){
                pstmt.setString(i + 1, param[i]);  //为预编译SQL设置参数
            }
        }
        num = pstmt.executeUpdate();                     //执行SQL语句
    } catch (SQLException e){
        e.printStackTrace();                              //处理SQLException异常
    } finally{
        //关闭当前的连接对象
        DbUtils.closeQuietly(conn, pstmt, null); //释放资源
    }
    return num;
}

/**
 * 得到所有的记录 ，需在外面关闭连接对象
 *
 * @param tableName   String 表名
 * @return ResultSet 结果集
 * @throws Exception
 */
publicResultSet getAllRecord(String tableName){
    try{
```

```
            conn = getConnection();
            pstmt = conn.prepareStatement("select * from " + tableName);
            rs = pstmt.executeQuery();
        } catch (Exception ex){
            ex.printStackTrace();
        }
        return rs;
    }

}
```

（3）在数据访问层建立数据访问层 UserDAO.java 接口类。

```
package com.ssmbook2020.dao;
import java.util.List;
import com.ssmbook2020.bean.User;
public interface UserDAO {
    //得到所有的用户信息
    public List<User> getAllUser();
    //添加用户
    public int saveUser(User user);
    //更新用户
    public int updateUser(User user);
    //根据条件，查找某个用户
    public User getUserByCondition(User user);
    //根据条件，查找所有符合条件的用户
    public  List<User> getSomeUserByCondition(User user);
}
```

这个接口比较简单，主要定义了访问数据库中需要的一些方法，如查找所有用户、保存用户、更新用户和根据一定条件查找用户等。

这里为什么定义接口而不是定义类？这是为后面引入 Spring 框架所做的铺垫。Spring 框架强调面向接口而不是面向类编程，引入 Spring 框架后读者将体会到这种做法的作用和优势。

（4）定义接口的实现类，类名为 UserDaoImpl.java。

```
package com.ssmbook2020.dao;
//省略导包代码
public class UserDaoImpl extends BaseDAOMySQL implements UserDAO {

    @Override
    public List<User>getAllUser(){
        String sql="select * from tbUser";
        List<User> userList =new ArrayList<User>();
        try{
            conn =super.getConnection();
            pstmt = conn.prepareStatement(sql);
```

```
            rs = pstmt.executeQuery(sql);
            while (rs.next()){
                int  userid = rs.getInt("userid"); //
                String username = rs.getString("username");
                String password="";         //不查询密码
                String email = rs.getString("email");
                String phone = rs.getString("phone");
                User user =new User(userid,username,password,email,phone);
                userList.add(user);
            }
        } catch (SQLException e){
            e.printStackTrace();
        } finally{
            DbUtils.closeQuietly(conn, pstmt, rs);
        }
          return userList;
    }

    @Override
    public intsaveUser(User user){
        //TODO Auto-generated method stub
        return 0;
    }

    @Override
    public intupdateUser(User user){
        //TODO Auto-generated method stub
        return 0;
    }

    @Override
    public UsergetUserByCondition(User user){
        //TODO Auto-generated method stub
        return null;
    }

    @Override
    public List<User>getSomeUserByCondition(User user){
        //TODO Auto-generated method stub
        return null;
    }
}
```

1.3.5 JUnit 测试框架

因为测试的需要，这里先完成 public List<User> getAllUser()这个方法，它的主要作用是查询数据库 tbuser 表中的所有用户记录。

完成这段代码之后，有以下两个疑问要解决，数据库的连接是否正确？能否正确查出 tbuser 表当中的数据？为此需要进行单元测试，而 Java 最常用的单元测试框架是 JUnit。

首先在项目中引入 JUnit 测试框架库，根据图 1-9 和图 1-10 的步骤标识在项目中加入 JUnit 包，注意选择的是 JUnit 4 版本。

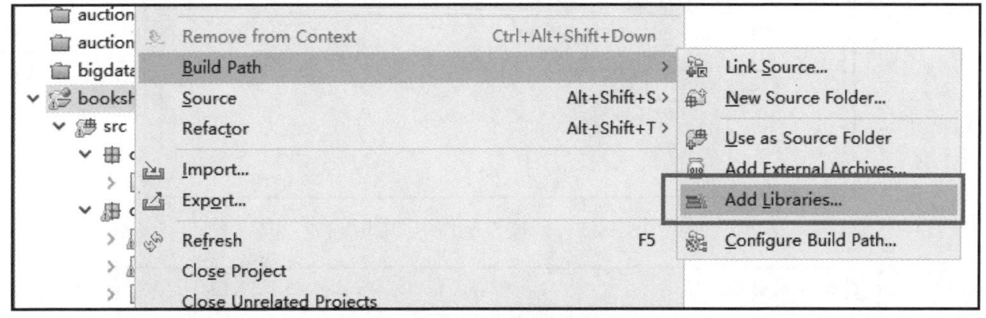

图 1-9　项目加入 Library

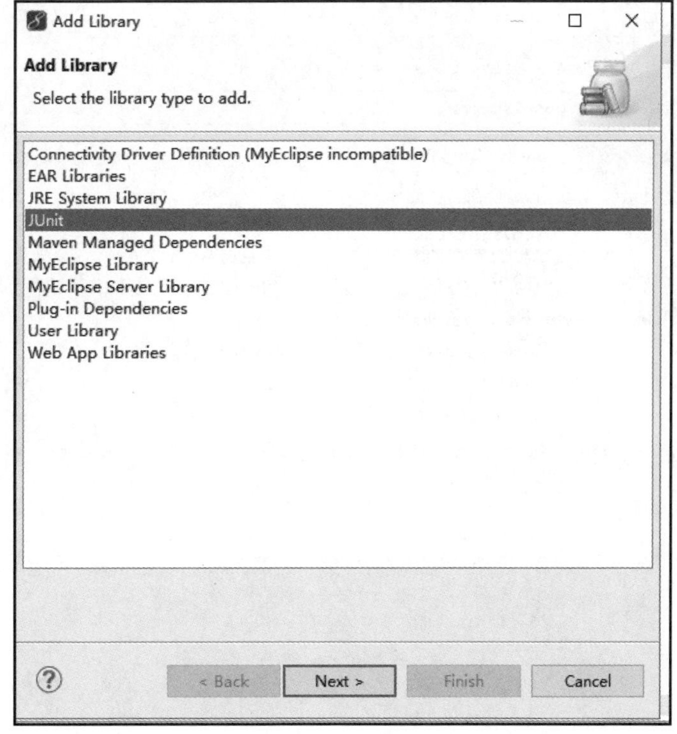

图 1-10　加入 JUnit

框架引入成功之后，接下来对 UserDaoImpl 类进行测试。需要新建一个测试用例，如图 1-11 所示，根据 MyEclipse 中的向导创建测试用例。之前引入的测试框架是 JUnit4 版本，所以这里的测试用例也要选择 JUnit 4 版本，它支持测试有关的注解。如图 1-12 所示，建立一个 UserDaoTestCase 测试类。图 1-13 表示要对目标类的哪些方法生成测试方法。

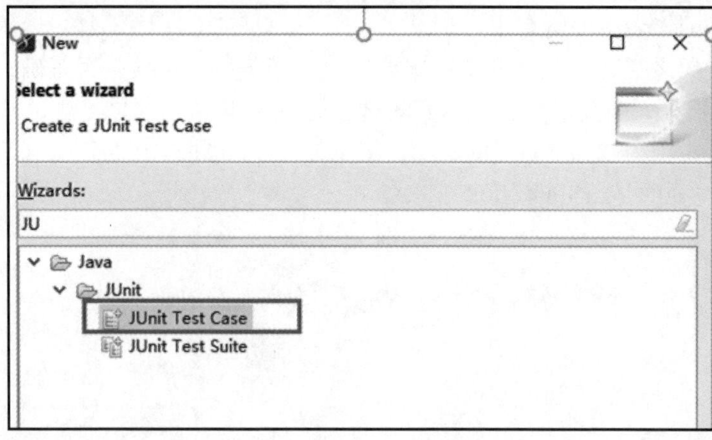

图 1-11　创建 JUnit 测试用例

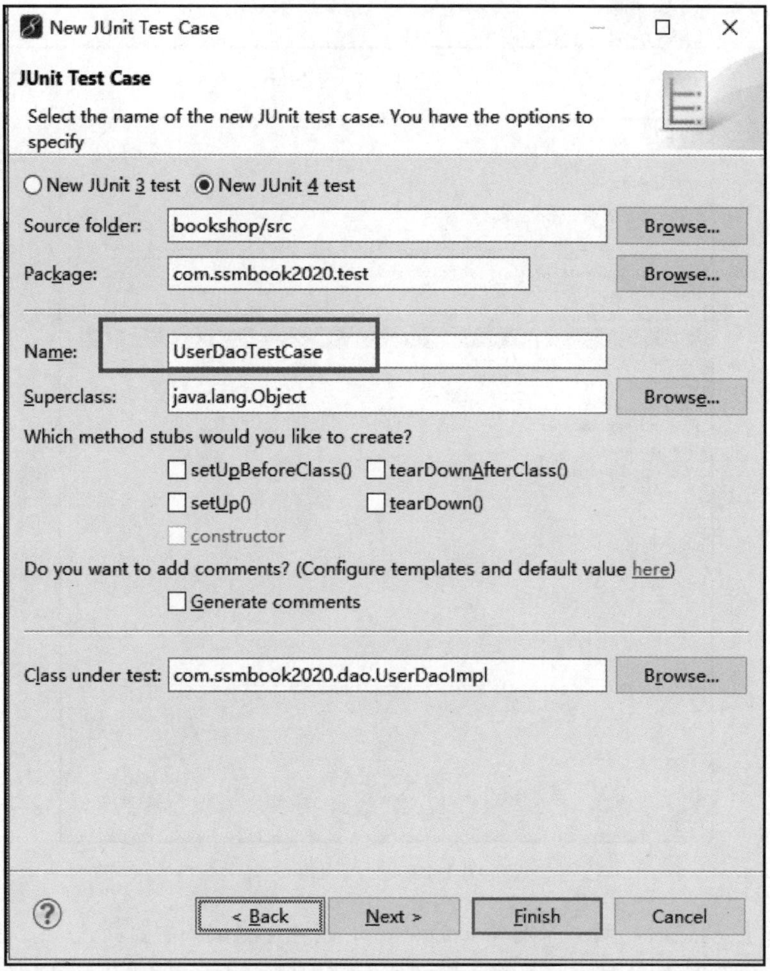

图 1-12　UserDaoTestCase 类

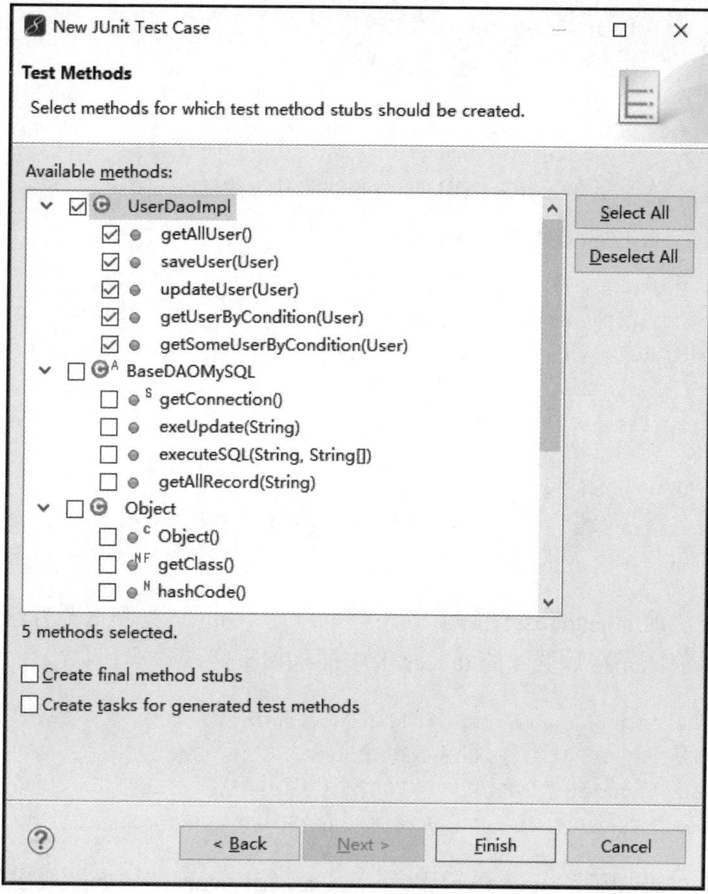

图 1-13　选择测试的目标方法

最后得到的方法代码如下。

```
package com.ssmbook2020.test;

import com.ssmbook2020.bean.User;
import com.ssmbook2020.dao.UserDaoImpl;

public class UserDaoTestCase {
    public voidtestGetAllUser(){
    fail("Not yet implemented");
    }
    public voidtestSaveUser(){
            fail("Not yet implemented");
    }
    public voidtestUpdateUser(){
        fail("Not yet implemented");
    }
    public voidtestGetUserByCondition(){
        fail("Not yet implemented");
    }
```

```
    public voidtestGetSomeUserByCondition(){
        fail("Not yet implemented");
    }
}
```

要完成测试，必须要在 testGetAllUser()方法当中加入如下代码。

```
    @Test
    public voidtestGetAllUser(){
        UserDaoImpl userdao =new UserDaoImpl();
        List<User> list =userdao.getAllUser();
        for (Useruser : list){
            System.out.println(user.toString());
        }

        assertEquals(2,list.size() );//判断数据是否有两条
    }
```

注意这个方法前面的@Test 注解，该注解可以让 JUnit 运行测试此方法。

在执行测试前，先在数据库的 tbuser 表中插入如下两条数据。

```
INSERT INTO 'tbuser' VALUES ('1001', 'Jack',
'*23AE809DDACAF96AF0FD78ED04B6A265E05AA257', ' Jack@qq.com', '15818119999');
INSERT INTO 'tbuser' VALUES ('1002', 'Rose',
'*23AE809DDACAF96AF0FD78ED04B6A265E05AA257', 'Rose@qq.com', '15818119997');
```

由于目前表中添加了两条记录，所以使用了 assertEquals 方法判断取出的结果是否等于两条。assertEquals 方法用于表示断言期望值和实际值是否相等，如果相等则成功，反之则失败。

图 1-14 为测试的运行结果。在图 1-14 中，第 1 个标注 Runs: 1/1 表示运行了一个方法，成功了一个。第 2 个标注表示测试的方法名和消耗时间。第 3 个标注是 JUnit 对测试结果的反馈，测试成功时是绿色的，测试失败时是红色的。如果测试多个方法，只有全部方法测试成功时才是绿色的，其中有任何一个失败都会是红色的状态条。第 4 个标注表示控制台打印结果。

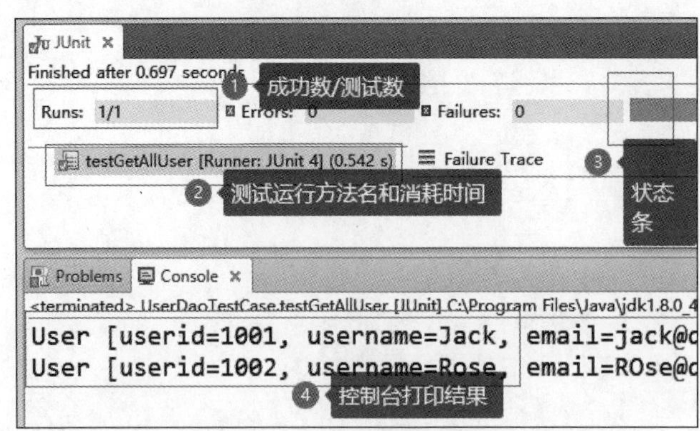

图 1-14 测试运行结果

以上针对数据访问层的测试通过后，说明已经成功连通数据库了，下面可以进行增、删、改等其他操作。

1.3.6 三层架构业务逻辑层实现

完成了数据访问层的代码后，接下来需要实现业务逻辑层代码。业务逻辑层要实现查询所有的用户信息，在数据访问层已经从数据库中读取了信息，在此只需要调用数据访问层对象即可。业务逻辑层的实现同样要遵循面向接口编程的思想，因此要先创建好业务层的接口。

业务逻辑层的接口类 UserService.java 的代码如下所示。

```
package com.ssmbook2020.service;
import java.util.List;
import com.ssmbook2020.bean.User;
public interface UserService {
      //得到所有的用户信息
       public List<User> getAllUser();
       //添加用户
       public int saveUser(User user);
       //更新用户
       public int updateUser(User user);
       //根据条件，查找某个用户
       public User getUserByCondition(User user);
       //根据条件，查找所有符合条件用户
       public  List<User> getSomeUserByCondition(User user);

}
```

该接口类几乎和数据访问层的接口类是一样的。这是由于此处没有特殊的业务要处理，所以就采用了和数据访问层完全一样的方法列表。

定义好接口后完成它的实现类 UserServiceImpl.java，其代码如下所示。

```
package com.ssmbook2020.service;
import java.util.List;
import com.ssmbook2020.bean.User;
import com.ssmbook2020.dao.UserDAO;
import com.ssmbook2020.dao.UserDaoImpl;
public class UserServiceImpl implements UserService {
   private UserDAO userdao =new UserDaoImpl();
   @Override
   public List<User>getAllUser(){
      return userdao.getAllUser();
   }
   publicUserDAO getUserdao(){
      return userdao;
   }
```

```
    public voidsetUserdao(UserDAO userdao){
        this.userdao = userdao;
    }
    @Override
    public intsaveUser(User user){
        //TODO Auto-generated method stub
        return 0;
    }
    @Override
    public intupdateUser(User user){
        //TODO Auto-generated method stub
        return 0;
    }
    @Override
    public UsergetUserByCondition(User user){
        //TODO Auto-generated method stub
        return null;
    }
    @Override
    public List<User>getSomeUserByCondition(User user){
        //TODO Auto-generated method stub
        return null;
    }
}
```

这里需要特别注意业务逻辑层对数据访问层的依赖关系，如图 1-15 所示的依赖关系标记。

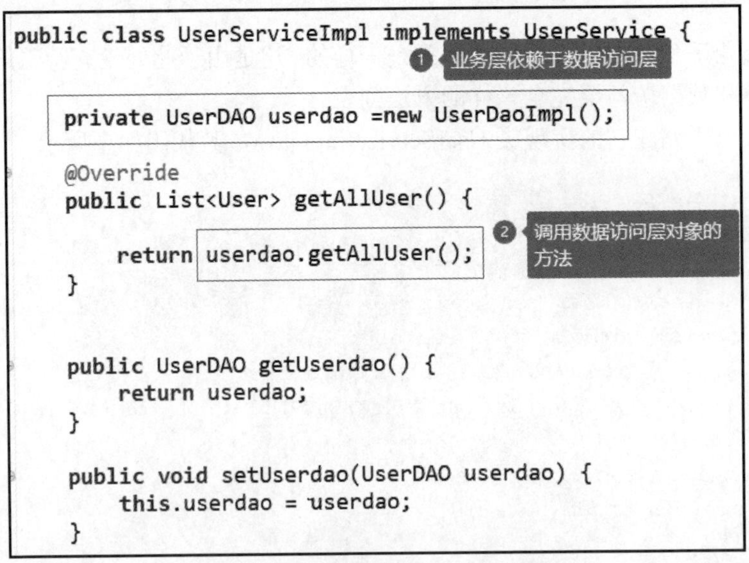

```
public class UserServiceImpl implements UserService {
                                  ❶ 业务层依赖于数据访问层
    private UserDAO userdao =new UserDaoImpl();

    @Override
    public List<User> getAllUser() {
                                              ❷ 调用数据访问层对象的
        return userdao.getAllUser();             方法
    }

    public UserDAO getUserdao() {
        return userdao;
    }

    public void setUserdao(UserDAO userdao) {
        this.userdao = userdao;
    }
```

图 1-15　业务层对数据访问层的依赖

1.3.7 三层架构 Web 层实现

Web 层主要用来和用户进行交互，对于浏览器/服务器架构的程序来说，就是用户发出请求、服务器响应请求。未采用任何框架的传统的 JSP Web 应用程序主要是基于 JSP+Servlet 组成的。下面实现用户信息的展示。

（1）新建一个 Servlet，其名为 UserServlet.java，其代码如下。

```
    package com.ssmbook2020.web;

    import java.io.IOException;
    import java.io.PrintWriter;

    import javax.servlet.ServletException;
    import javax.servlet.http.HttpServlet;
    import javax.servlet.http.HttpServletRequest;
    import javax.servlet.http.HttpServletResponse;

    public class UserServlet extends HttpServlet{

        private void handlerRequest(HttpServletRequest request,
                HttpServletResponse response) throws ServletException,
IOException{
            //暂时为空
        }
        public voiddoGet(HttpServletRequest request, HttpServletResponse
response) throws ServletException, IOException{
             handlerRequest(request,response);
        }
        public voiddoPost(HttpServletRequest request, HttpServletResponse
response) throws ServletException, IOException{
         handlerRequest(request,response);
        }

    }
```

为了让用户发出的 get 或 post 请求都采用同样的处理方式，在这个类中创建了一个名为 handlerRequest()的方法，该方法暂时为空。

（2）利用向导新建 Servlet 时，MyEclipse 会在 web.xml 文件中自动配置好该 Servlet 的访问路径，其代码如下。

```
    <?xml version="1.0" encoding="UTF-8"?>
    <web-app xmlns:xsi="http://www.w3.org/2001/XMLSchema-instance" xmlns=
"http://java.sun.com/xml/ns/javaee"
    xsi:schemaLocation="http://java.sun.com/xml/ns/javaee
    http://java.sun.com/xml/ns/javaee/web-app_3_0.xsd"id="WebApp_ID" version=
"3.0">
      <display-name>bookshop</display-name>
```

```
  <servlet>
    <servlet-name>UserServlet</servlet-name>
    <servlet-class>com.ssmbook2020.web.UserServlet</servlet-class>
  </servlet>

  <servlet-mapping>
    <servlet-name>UserServlet</servlet-name>
    <url-pattern>/servlet/UserServlet</url-pattern>
  </servlet-mapping>
  <welcome-file-list>
    <welcome-file>index.jsp</welcome-file>
</web-app>
```

其中，<url-pattern>/servlet/UserServlet</url-pattern>配置用户访问该 Servlet 的地址。

为了让 UserServlet 工作起来，必须要引用业务逻辑层的类和方法，添加代码如下。

```
package com.ssmbook2020.web;
//省略导包的 import 语句
public class UserServlet extends HttpServlet {

    UserService userService =new UserServiceImpl();

    publicUserService getUserService(){
        return userService;
    }

    public voidsetUserService(UserService userService){
        this.userService = userService;
    }

}
```

然后之前定义的 handlerRequest 方法就可以利用 userService 这个对象，该方法的代码及说明如图 1-16 所示。

```
private void handlerRequest(HttpServletRequest request,
        HttpServletResponse response) throws ServletException, IOException {
    //约定请求的地址使用action作为参数名
    String action =request.getParameter("action");
    if(action.equalsIgnoreCase("listUser")){
          //调用业务逻辑层的方法，得到用户列表数据
        List<User> userList =userService.getAllUser();
        //将数据放入Request域中，以便JSP页面可以访问它
        request.setAttribute("userList", userList);
        request.getRequestDispatcher("/listUser.jsp").forward(request, response);
    }else{
        response.sendRedirect("index.jsp");
    }
}
```

❶ 请求地址为servlet/UserServlet?action=listUser

❷ 将数据传入Request域中

图 1-16　handlerRequest 处理用户请求

这里约定要读取所有的用户信息时发出的请求地址为 http://localhost:8080/bookshop/

servlet/ UserServlet?action=listUser，展示用户信息的页面为 listUser.jsp 页面。

（3）建立一个静态的页面用来显示用户信息，其代码如下。

```
<%@ page language="java" import="java.util.*" pageEncoding="UTF-8"%>
<!DOCTYPE HTML PUBLIC "-//W3C//DTD HTML 4.01 Transitional//EN">
<html>
  <head>
    <title>显示用户信息</title>
  </head>
  <body>
  <div align="center">
   用户人数 : <br/>
   信息列表如下:
   <table border="1" >
     <thead>
      <tr>
          <td>用户 ID</td>
          <td>用户名</td>
          <td>用户 email</td>
          <td>用户电话</td>
      </tr>
     </thead>
       <tbody>
      <tr>
          <td>1</td>
          <td>jack</td>
          <td>test@qq.com</td>
          <td>15815810001</td>
      </tr>
     </tbody>
   </table>
   </div>
  </body>
</html>
```

将此项目部署到 Tomcat 服务器上，通过浏览器访问 http://localhost:8080/bookshop/listUser.jsp，得到如图 1-17 所示的页面。

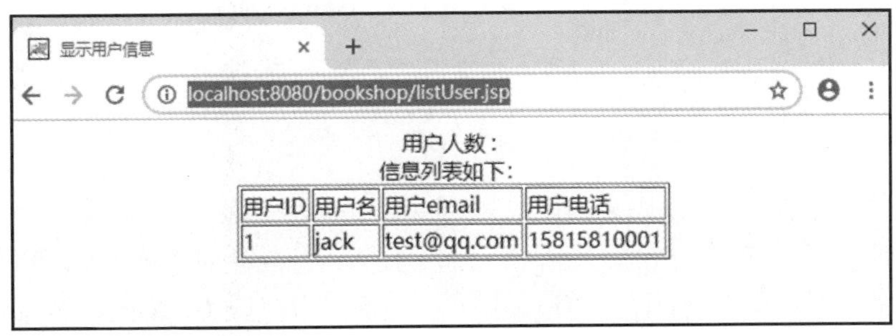

图 1-17　静态结果

要能正确地得到并显示 Servlet 放在 Request 域中的数据，并且不在页面中嵌入 Java 代码，必须要借助于 JSTL 标签库才行。引入 JSTL 标签库的代码如图 1-18 所示。

```
<%@ page language="java" import="java.util.*" pageEncoding="UTF-8"%>
<%@page import="com.ssmbook2020.bean.User"  %>
<!-- 导入jstl的标准核心库 -->
<%@taglib uri="http://java.sun.com/jsp/jstl/core" prefix="c" %>
<!-- 导入jstl的函数库 -->
<%@taglib uri="http://java.sun.com/jsp/jstl/functions" prefix="fn" %>
<!DOCTYPE HTML PUBLIC "-//W3C//DTD HTML 4.01 Transitional//EN">
<html>
```

① 导入User类
② 导入JSTL核心库
③ 导入JSTL函数库

图 1-18　引入 JSTL 标签库

在 listUser.jsp 页面中，通过 JSTL 的函数标签库显示集合中的数据数目，通过核心标签库循环读取 List 中的数据。使用标签库的代码如图 1-19 所示。

```
用户人数 :${fn:length(requestScope.userList) } <br/>
信息列表如下：
  <table border="1" >
    <thead>
     <tr>
         <td>用户ID</td><td>用户名</td>
         <td>用户email</td><td>用户电话</td>
     </tr>
    </thead>
      <tbody>
     <c:forEach items="${requestScope.userList }" var="user">
         <tr>
             <td>${user.userid}</td><td>${user.username}</td>
             <td>${user.email}</td><td>${user.phone}</td>
         </tr>
     </c:forEach>
```

① 函数标签库，显示集合中对象个数
② 核心标签库，循环读取集合数据

图 1-19　使用标签库

然后必须通过访问 Servlet 跳转到该页面才可以使用数据。访问地址为 http://localhost:8080/bookshop/servlet/UserServlet?action=listUser ，得到的结果如图 1-20 所示。

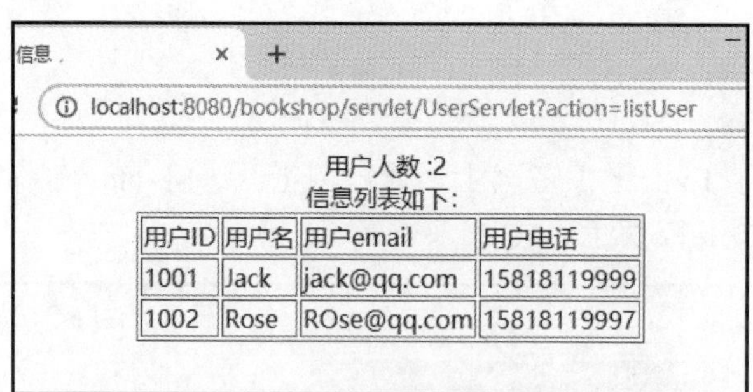

图 1-20　动态显示用户表页面

至此，已经完成了一个会员模块中显示用户的功能点。

为了方便使用，可以在首页 index.jsp 页面中加入访问功能的链接地址。用户管理模块除已经完成的显示用户功能外，还有添加用户、更新用户信息（如修改、删除）等。

本书后续的章节将会继续完成此项目。

1.4 本章小结

本章首先介绍了 Spring 框架的作用，然后介绍了要安装的软件环境，接着通过 GoodBook 网上书店的案例实现了用户显示模块，最后讲解了基于 Servlet+JDBC+JSP+JSTL/EL 三层架构的开发过程。使用编程开发框架可以减少工作量并提高开发效率。在 Web 层可以引入 Spring MVC 来替代使用原生 Servlet，在数据访问层可以引入 Spring JDBC 或 MyBatis 框架来将开发者从原生 JDBC 的烦琐代码中解放出来，在业务层可以通过丰富的工具集来解决事务管理、电子邮件发送、缓存管理等问题。

习 题 1

1. 在本章案例基础上实现添加用户功能。
2. 在本章案例基础上实现更改用户信息功能。

第2章 Spring IoC

本章学习内容

- Spring 框架介绍；
- IoC 容器的入门；
- Bean 标签的属性配置；
- 对象之间的依赖注入；
- 使用注解的方式来管理对象。

2.1 Spring 框架简介

在学习 Spring 框架之前，有必要先了解一下该软件的创建者 Rod Johnson，如图 2-1 所示。他拥有悉尼大学计算机科学学位，更令人吃惊的是，他在进入软件开发领域前还获得了音乐学的博士学位。他有着丰富的 C/C++技术背景，在 1996 年就开始了对 Java 服务器端技术的研究，或许只有这样的人才能设计出如此有艺术气息的 Spring 框架。

图 2-1　Rod Johnson

Spring 框架有如下几个特点。

- Spring 是分层的 Java SE/EE 应用的全栈型轻量级开源框架。Spring 不但可以用于 Java SE 的开发，而且可以用于 Java EE 的企业级开发。所谓全栈型是指在企业开发的表示层、业务层和持久层都提供了技术解决方案，它提供了表示层 SpringMVC、持久层 Spring JDBC 和业务层的事务管理等企业级应用解决方案。同时 Spring 是一个免费开源的轻量级框架，被广大开发者所使用。
- Spring 以控制反转（Inversion of Control, IoC）和面向切面编程（Aspect-Oriented Programming, AOP）为核心。
- Spring 将开源世界中众多优秀的第三方框架和类库整合到 Spring 框架中，让其他优秀的框架与之一起成长，所以 Spring 框架成为越来越受欢迎的 Java EE 企业级应用框架。

Spring 的 IoC 和 AOP 是它的两个核心，在此进行简要介绍，后面将使用两章详细介绍。

1. 控制反转（IoC）

Spring 通过一种称为控制反转的技术促进了松耦合。在使用了 IoC 后，一个对象依赖的其他对象会通过被动的方式传递进来，而不是这个对象自己创建或者查找依赖对象。也就是说，不是对象从容器中查找依赖，而是容器在对象初始化时不等对象请求就主动将依赖传递给它。

以前主动创建对象的方式如图 2-2 所示。

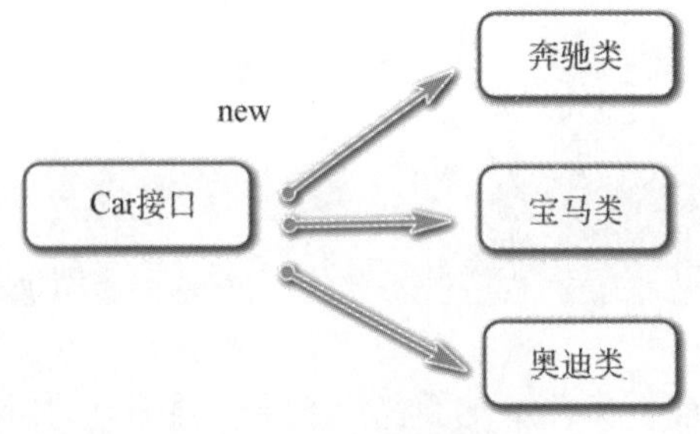

图 2-2　创建对象

```
//Car 是接口，后面是实现类
Car bmw = new Bmw();
Car benZ = new BenZ();
Car audi = new Audi();
```

现在被动接收工厂创建好的对象，从容器中获取对象来使用，并且由容器管理对象之间的依赖关系，如图 2-3 所示。

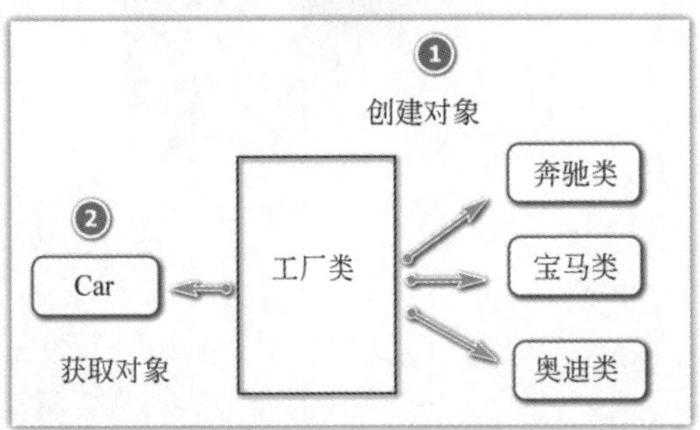

图 2-3　控制反转

```
Car bmw = BeanFactory.getBean("bmw");
Car benZ = BeanFactory.getBean("benZ");
Car audi = BeanFactory.getBean("audi");
```

使对象的控制权发生了翻转，所以称为控制反转（IoC）。

2. 面向切面编程（AOP）

Spring 提供了面向切面编程的丰富支持，允许通过分离应用的业务逻辑与其他系统级服务进行开发。应用对象只实现它们的业务逻辑就可以了，与应用无关但又必需的一些代码，如日志记录、事务处理、错误处理等功能，可以写在另外一个地方，由 Spring 把它们组合在一起运行以实现相应的功能。这个知识点在下一章再详细介绍。

2.2 Spring 的体系结构

视频讲解

Spring 框架集成了 20 多个模块，图 2-4 为 Spring 的体系结构图。

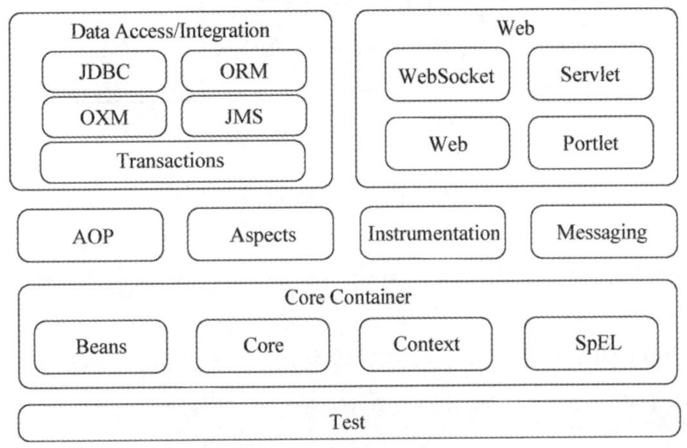

图 2-4 Spring 的体系结构

（1）核心容器模块

- Beans 模块：提供 BeanFactory 工厂模式的实现，Spring 将所有管理对象称为 Bean。
- Core 模块：提供 Spring 框架的基本组成部分，包括 IoC 和依赖注入（Dependency Injection, DI）的功能。
- Context 模块：访问和配置 Bean 对象的上下文对象，即核心容器对象，如 ApplicationContext。

（2）数据访问与集成模块

- JDBC 模块：提供 JDBC 访问数据库的支持，如 JdbcTemplate。
- ORM 模块：对主流的 ORM 框架提供支持，如 JPA、JDO、Hibernate 等。
- Transactions 模块：支持声明式事务的管理。

（3）Web 模块

- Servlet 模块：包含 SpringMVC 和 REST Web Services 实现的 Web 应用程序。
- Web 模块：提供基本的 Web 开发集成特性，如文件上传、收发邮件等。

（4）其他模块

- AOP 模块：提供面向切面编程的功能。
- Aspects 模块：提供与 AspectJ 框架集成的功能。
- Test 模块：提供对单元测试和集成测试的支持。

2.2.1 Spring 环境简介

Spring 的官网地址为 https://spring.io/，从这里可以获取最新的版本和开发文档，如图 2-5 所示。

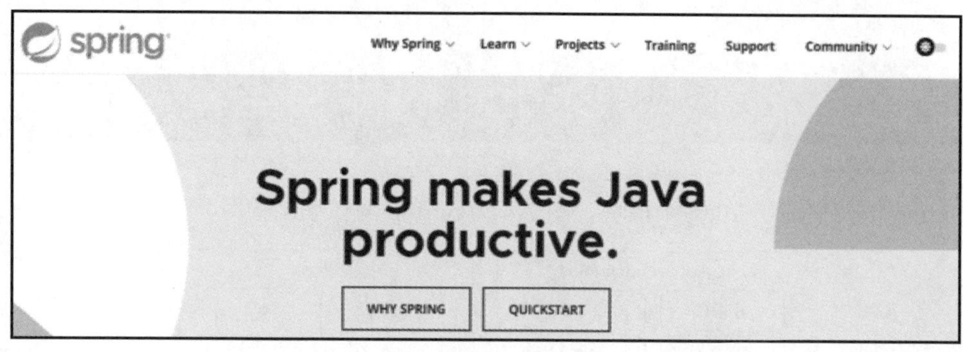

图 2-5　Spring 官网

通常 MyEclipse 开发工具自带常用的开发框架包，无须手动下载。本书使用的 MyEclipse2019 自带的最高版本是 Spring 4.1.0，如果希望下载最新的包，可以在 https://repo.springsource.org/libs-release-local/org/springframework/spring/ 中下载。Spring 框架的版本列表如图 2-6 所示。

例如，下载 5.2.0.RELEASE 版本，本地下载得到的是名为 spring-5.2.0.RELEASE-dist.zip 的压缩文件，解压后的目录结构如图 2-7 所示。

4.3.6.RELEASE/	30-Sep-2019 11:06
4.3.7.RELEASE/	30-Sep-2019 11:06
4.3.8.RELEASE/	30-Sep-2019 11:06
4.3.9.RELEASE/	30-Sep-2019 11:06
5.0.0.RELEASE/	30-Sep-2019 11:06
5.0.1.RELEASE/	30-Sep-2019 11:06
5.0.10.RELEASE/	30-Sep-2019 11:06
5.0.11.RELEASE/	30-Sep-2019 11:06
5.0.12.RELEASE/	30-Sep-2019 11:06
5.0.13.RELEASE/	30-Sep-2019 11:06
5.0.14.RELEASE/	30-Sep-2019 11:06
5.0.15.RELEASE/	30-Sep-2019 11:06

图 2-6　Spring 框架版本列表

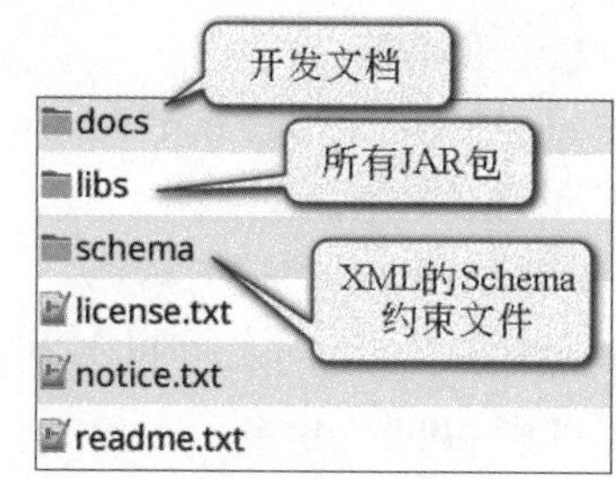

图 2-7　Spring 压缩包的目录结构

后续如果使用 Maven 工具来管理工程的 JAR 包，可以简化包的管理。

2.2.2　IoC 容器入门

在 Spring 框架的容器继承树中，最顶层的接口是 BeanFactory，这是工厂类的一种实现。常用的子接口是 ApplicationContext，常用的实现类有以下三个。

（1）从类路径（即 src 目录）下读取配置文件，创建容器类 ClassPathXmlApplicationContext。

（2）从本地配置文件中使用绝对地址的方式读取配置文件，创建容器类 ClassPathXmlApplicationContext。

（3）从配置类中使用注解的方式读取配置信息，创建容器类 AnnotationConfigApplicationContext。

本书主要使用 ClassPathXmlApplicationContext 和 AnnotationConfigApplicationContext 这两个容器。Spring 的容器类的结构关系如图 2-8 所示。

图 2-8　Spring 的容器类

下面通过示例对 Spring IoC 容器进行分析。

1. 采用传统方式创建对象

（1）创建新的 Java 工程。创建工程页面如图 2-9 所示，工程结构如图 2-10 所示。

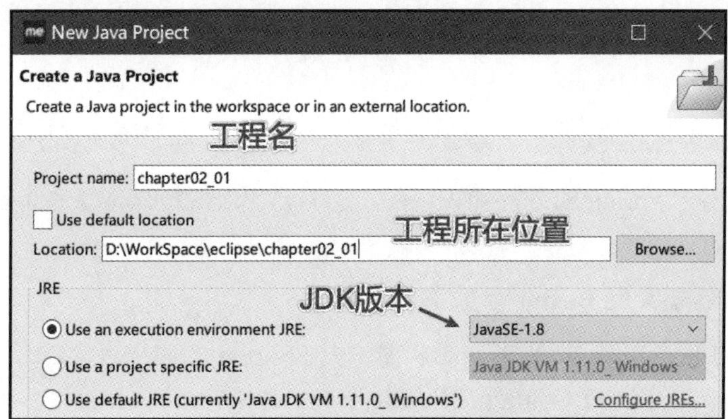

图 2-9　创建新的 Java 工程

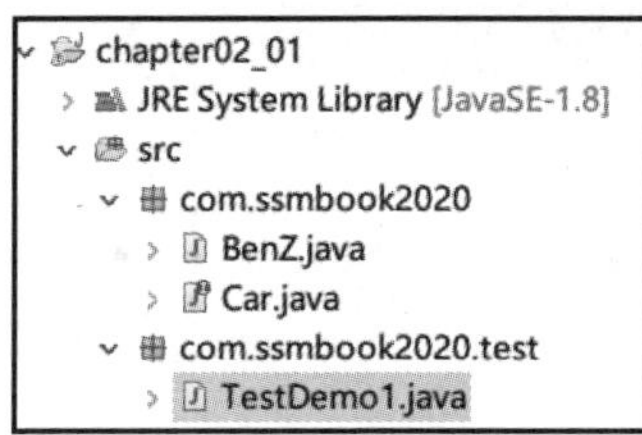

图 2-10　工程结构

（2）创建接口 Car。

```
package com.ssmbook2020;
public interface Car {
    void drive();
}
```

（3）创建实现类 Benz。

```
package com.ssmbook2020;
public class BenZ implements Car {
    @Override
    public void drive(){
        System.out.println("开奔驰车");
    }
}
```

（4）创建测试类。

```
package com.ssmbook2020.test;
import com.ssmbook2020.BenZ;
import com.ssmbook2020.Car;
public class TestDemo1 {
    public static void main(String[] args){
        Car benZ = new BenZ();   //主动创建对象
        benZ.drive();
    }
}
```

（5）运行结果。

```
开奔驰车
```

该案例没有用到 Spring 的任何知识，有一定 Java 基础即可看懂。下面添加 Spring 容器进行管理。

2. 使用 IoC 容器装配 Bean

（1）将所有的 Java Bean 放入 Spring 容器中，由 Spring 实例化对象。如果使用 MyEclipse 开发工具，则选择 Project → Configure Facets → Install Spring Facet 命令，系统会自动在 src 目录下建立一个 applicationContext.xml 文件（默认配置文件名），项目结构如图 2-11 所示。

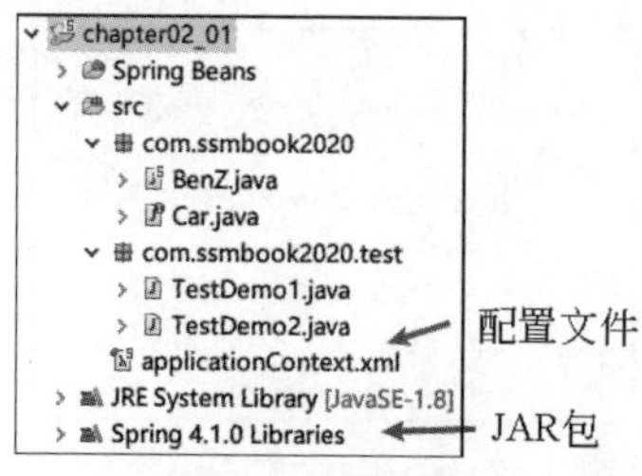

图 2-11　项目结构

（2）在 applicationContext.xml 文件中配置奔驰汽车对象。

```
<?xml version="1.0" encoding="UTF-8"?>
<beans
    xmlns="http://www.springframework.org/schema/beans"
    xmlns:xsi="http://www.w3.org/2001/XMLSchema-instance"
```

```
    xmlns:p="http://www.springframework.org/schema/p"
    xsi:schemaLocation="http://www.springframework.org/schema/beans
    http://www.springframework.org/schema/beans/spring-beans-4.1.xsd">
    <!-- 奔驰车类 -->
    <bean class="com.ssmbook2020.BenZ" id="benZ"/>
</beans>
```

上面配置中的<bean>标签代表放入 Spring 容器中的 Java Bean 对象，所有的类现在都由 Spring 来管理，id 可以理解为对象名，class 则是完全限定类名（即包含包名的类名）。

（3）修改测试类，代码如下。

```
package com.ssmbook2020.test;
import org.springframework.context.support.ClassPathXmlApplicationContext;
import com.ssmbook2020.Car;
public class TestDemo2 {
    public static void main(String[] args){
        //创建 Spring 容器，默认是从类路径下读取 applicationContext.xml 配置文件
        ClassPathXmlApplicationContext context = new
ClassPathXmlApplicationContext("applicationContext.xml");
        //从容器中获取汽车对象
        Car benZ = (Car) context.getBean("benZ");
        //调用汽车类的方法
        benZ.drive();
        //关闭容器
        context.close();
    }
}
```

（4）运行测试类会发现前面多出几句红色的警告，这是因为 Spring 容器使用 log4j 作为日志记录组件，log4j 需要相应的配置文件。在项目中加入 log4j.properties，将其放在 src 目录下就可以避免这些警告了。

但要注意两点：①文件要放在 src 根目录下；②文件名必须是 log4j.properties。

log4j.properties 是一个特殊格式的文本文件，该文件的内容如下。

```
#显示在控制台上#
log4j.appender.stdout=org.apache.log4j.ConsoleAppender
log4j.appender.stdout.Target=System.out
log4j.appender.stdout.layout=org.apache.log4j.SimpleLayout
#日志输出级别：fatal/error/warn/info/debug#
log4j.rootLogger=info stdout
```

两次测试代码运行的效果是一样的，一个是按以前的主动方式创建对象，一个是由 Spring IoC 容器来管理对象并从容器中获取对象。

2.2.3 Bean 标签的配置

Bean 标签的作用是在 Spring 容器中创建一个对象。表 2-1 是 Bean 的属性说明。

表 2-1　Bean 的属性说明

属　　性	说　　明
id	放在容器中的唯一标识
name	对象可以有多个名字，多个名字之间使用逗号、空格、分号隔开都可以，也可以通过名字从容器中获取对象
class	指定类全名，需要指定的是实现类而不是接口，因为接口不能被实例化
scope	指定 Bean 在容器中的作用范围，常用的有以下两个取值。 singleton：默认值，单例对象，整个容器中只会创建一个对象。 prototype：多例对象，每次都会从容器中获取一个新的对象。 该属性还有其他取值，但在企业开发过程中使用不多，所以本书未提及
lazy-init	是否使用延迟加载，默认情况下容器一创建就同时创建容器中的所有对象
init-method	创建对象时指定执行的初始化方法名
destroy-method	销毁对象时指定执行的销毁方法名

其中，scope 属性表示对象作用域范围。以前 Bean 只有两种作用域，即 singleton（单例）、non-singleton（也称 prototype），后来增加了 session、request、global session 这 3 种专用于 Web 应用程序上下文的 Bean。

当 Bean 的作用域设置为 singleton（默认值）时，Spring IoC 容器中只会存在一个共享的 Bean 实例。也就是说，当把 Bean 的作用域设置为 singleton 时，Spring IoC 容器只会创建该 Bean 定义的唯一实例。这个单一实例会被存储到单例缓存中，并且所有针对该 Bean 的后续请求和引用都将返回被缓存的对象实例。

prototype 作用域部署的 Bean，每次请求（将其注入另一个 Bean 中，或者以程序的方式调用容器的 getBean()方法）都会产生一个新的 Bean 实例，相当于一个 new 的操作。Spring 容器不对 prototype Bean 的整个生命周期负责，容器在初始化、配置或装配完 prototype 实例后，将它交给客户端，随后就对该 prototype 实例不再理会了。清除 prototype 作用域的对象并释放任何 prototype Bean 所持有的资源，都需要用户自己写代码完成。

scope 属性还有以下 3 个作用域，仅在基于 Web 的 ApplicationContext 情形下有效。

（1）request：在一次 HTTP 请求中，容器会返回该 Bean 的同一个实例，而对于不同的用户请求会返回不同的实例。

（2）session：与 request 相同，唯一的区别是请求的作用域变成了 session。

（3）global session：在全局的 HTTP Session 中，容器会返回该 Bean 的同一个实例。这主要用于分布式开发中，同一个用户的会话可能会出现在不同的服务器上的情况。

2.2.4　scope 属性值

通过一个案例来演示两个 scope 属性值的区别。分别两次从容器中获取 benZ 对象，输出它们的对象值，因为没有重写 toString()方法，所以输出的是它的内存地址，由此比较是否为同一个对象。

（1）创建测试类，其代码如下。

```
package com.ssmbook2020.test;
import org.springframework.context.support.ClassPathXmlApplicationContext;
```

```
import com.ssmbook2020.Car;
public class TestDemo3 {
    public static void main(String[] args){
        //创建 Spring 容器，默认是从类路径下读取 applicationContext.xml 配置文件
        ClassPathXmlApplicationContext context = new
ClassPathXmlApplicationContext("applicationContext.xml");
        //从容器中获取汽车对象
        Car c1 = (Car) context.getBean("benZ");
        Car c2 = (Car) context.getBean("benZ");
        System.out.println("第 1 个对象:" + c1);
        System.out.println("第 2 个对象:" + c2);
        System.out.println(c1 == c2);
        //关闭容器
        context.close();
    }
}
```

运行结果如下。

```
第 1 个对象:com.ssmbook2020.BenZ@72cde7cc
第 2 个对象:com.ssmbook2020.BenZ@72cde7cc
true
```

由此可以发现，两次获取的是同一个对象，它们的地址相同。这说明在默认的情况下 scope 的取值是 singleton 单例对象。

（2）修改 applicationContext.xml 配置文件，将 scope 的取值换成 prototype。

```
<bean class="com.ssmbook2020.BenZ" id="benZ" scope="prototype"/>
```

（3）再次运行上面的测试类，可以发现，每次获取的对象都是不同的对象。

```
第 1 个对象:com.ssmbook2020.BenZ@2cd2a21f
第 2 个对象:com.ssmbook2020.BenZ@2e55dd0c
false
```

由此可以得到如表 2-2 所示的结论。

表 2-2　scope 属性的取值

scope 取值	作用范围	生命周期
singleton 单例对象	容器一创建就创建这个对象，只要容器不销毁就一直存在	出生：容器创建时 活着：只要容器没有销毁就一直存在 死亡：容器关闭时
prototype 多例对象	每次获取对象就创建一个新的对象，使用完毕会被 GC（Garbage Collector，垃圾回收器）回收	出生：获取对象时 活着：使用过程中 死亡：由 GC 在合适时回收

视频讲解

2.3 依赖注入

2.3.1 依赖注入简介

首先给出依赖注入的定义：依赖注入（Dependency Injection, DI）是 Spring 框架核心 IoC 的具体实现。在编写程序代码时，通过控制反转把对象交给 Spring 管理。代码中必然会存在一定的依赖关系，如在业务层（Service）中需要引用持久层（DAO）对象，这种层与层之间的关系也交给 Spring 来维护。简单来说，依赖注入就是给成员变量赋值。

当某个 Java 对象（调用者）需要调用另一个 Java 对象（被调用者，即被依赖对象）时，以前调用者通常采用“new 被调用者”的代码方式来创建对象，这种方式会导致调用者与被调用者之间的耦合性增加，不利于后期项目的升级和维护，如图 2-12 所示。

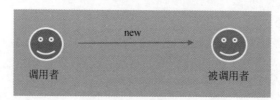

图 2-12　主动创建

在使用 Spring 框架后，对象的实例不再由调用者来创建，而是由 Spring 容器来创建，由容器控制程序之间的关系，而不是由调用者的程序代码直接控制。由容器负责将被依赖对象赋值给调用者的成员变量，相当于为调用者注入了它的依赖实例，这就是 Spring 的依赖注入，如图 2-13 所示。

图 2-13　依赖注入

2.3.2 构造器注入

构造器注入是指在类的构造方法实例化时，通过传递参数进行注入。

构造器注入的属性如表 2-3 所示。

表 2-3　构造器注入的属性

属性名	描　述
index	构造方法参数的位置，从 0 开始计数
name	构造方法中的参数名，与 index 属性两者选其一即可

续表

属性名	描　述
type	指定参数的类型，通常可以省略
value	用于注入简单类型，简单类型是指 String 类型和 8 种 Java 的基本类型
ref	用于注入引用类型

下面通过具体案例来学习构造器注入的方法。

（1）编写新的类 Customer 客户类，并添加以下属性。

```
package com.ssmbook2020;
import java.util.Date;
/**
 * 客户对象
 */
public class Customer {
    private int id;          //编号
    private String name;     //姓名
    private boolean male;    //性别是否为男
    private Date birthday;   //生日

    //无参构造方法
    public Customer(){
    }

    //全参构造方法
    public Customer(int id, String name, boolean male, Date birthday){
        this.id = id;
        this.name = name;
        this.male = male;
        this.birthday = birthday;
    }

    @Override
    public String toString(){
        return "Customer{" + "id=" + id +", name='" + name + '\'' + ", male="
+ male + ", birthday=" + birthday + '}';
    }
}
```

（2）修改 applicationContext.xml 配置文件，使用构造方法的方式注入所有的属性。

```
<bean class="com.ssmbook2020.Customer" id="customer">
    <!-- 注入简单类型  -->
    <constructor-arg name="id" value="100"/>
    <constructor-arg name="name" value="白骨精"/>
    <constructor-arg name="male" value="true"/>
    <!-- 注入引用类型  -->
```

```
        <constructor-arg name="birthday" ref="birthday" type="java.util.Date"/>
    </bean>
    <!-- 引用类型：现在的时间 -->
    <bean class="java.util.Date" id="birthday"/>
```

（3）在测试类中得到客户对象，输出客户对象。

```
    public class TestDemo5 {

        public static void main(String[] args){
            ClassPathXmlApplicationContext context = new
ClassPathXmlApplicationContext("applicationContext.xml");
            //从容器中获取对象
            Customer customer = (Customer) context.getBean("customer");
            System.out.println(customer);
            //关闭容器
            context.close();
        }
    }
```

（4）运行结果如下。

```
    Customer{id=100, name='白骨精', male=true, birthday=Mon Jan 18 17:13:07 CST
2021}
```

由此可以发现，通过构造方法注入了对象的所有属性，并且输出了属性值。

2.3.3 使用 set 注入

使用 set 注入是指通过类中的 set 方法给成员变量赋值，前提是对象要有 set 方法。set 注入的属性如表 2-4 所示。

表 2-4　set 注入的属性

属性名	属 性 说 明
name	要注入的属性名，本质上还是调用的 set 方法
value	要注入的属性值，用于简单类型的注入
ref	用于引用类型的注入

下面通过具体案例来学习使用 set 注入的方法。

（1）给 Customer 对象添加 set 方法。

```
    import java.util.Date;
    /**
     * 客户对象
     */
    public class Customer {
        private int id;            //编号
        private String name;       //姓名
```

```
    private boolean male;    //性别是否为男
    private Date birthday;   //生日

    public void setId(int id){
        this.id = id;
    }

    public void setName(String name){
        this.name = name;
    }

    public void setMale(boolean male){
        this.male = male;
    }

    public void setBirthday(Date birthday){
        this.birthday = birthday;
    }

     //省略了构造方法
    @Override
    public String toString(){
        return "Customer{" + "id=" + id +", name='" + name + '\'' + ", male="
+ male + ", birthday=" + birthday + '}';
    }
}
```

（2）修改 applicationContext.xml 配置文件，通过 property 元素给所有的属性赋值，将原来的构造方法注入部分的代码注释掉。

```
<!-- 使用 set 方法注入 -->
<bean class="com.ssmbook2020.Customer" id="customer">
    <!-- 注入简单类型和引用类型 -->
    <property name="id" value="200" />
    <property name="name" value="猪八戒" />
    <property name="male" value="true" />
    <property name="birthday" ref="birthday" />
</bean>

<!-- 引用类型：现在的时间 -->
<bean class="java.util.Date" id="birthday" />
```

（3）运行测试类，输出 Customer 对象的属性值。

```
Customer{id=200, name='猪八戒', male=true, birthday=Mon Jan 18 17:25:16 CST
2021}
```

由此可以发现，通过 set 方法注入并输出了所有的属性值。

2.3.4 使用 p 命名空间

在新版 Spring 中加入了使用 p 命名空间注入属性值，它的特点是使用<bean>的属性而不是以子元素的形式对属性注入，从而简化了配置代码。本质上还是使用 set 注入的方式，运行的效果也是一样的。p 命名空间的语法如下。

（1）对于简单类型（基本数据类型、字符串）的属性，使用语法 p:属性名="属性值"。

（2）对于引用类型的属性，使用语法 p:属性名-ref="Bean 的 id"。

使用该方法前要在配置文件中引入 p 命名空间，其语法格式为 xmlns:p="http://www.springframework.org/schema/p"。添加页面如图 2-14 所示。

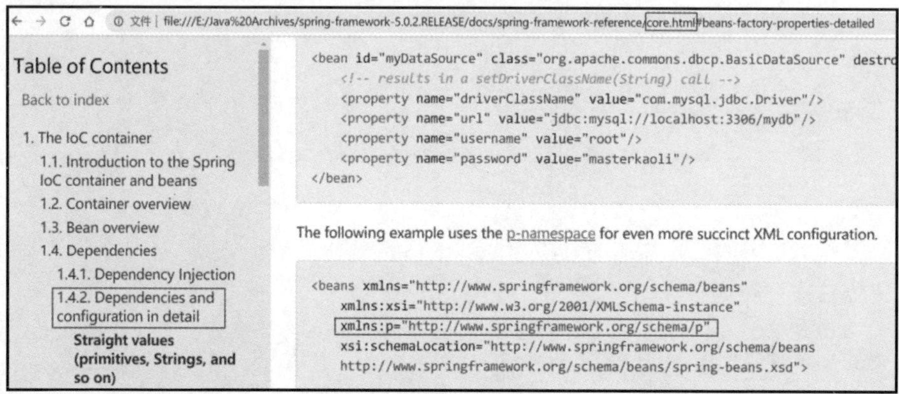

图 2-14　添加 p 命名空间

下面通过具体案例来学习使用 p 命令空间的方法。

（1）修改 applicationContext.xml 配置文件，将之前的 set 注入代码注释掉，使用 p 命名空间注入属性值。

```
<?xml version="1.0" encoding="UTF-8"?>
<beans xmlns="http://www.springframework.org/schema/beans"
    xmlns:xsi="http://www.w3.org/2001/XMLSchema-instance"
    xmlns:p="http://www.springframework.org/schema/p"
    xsi:schemaLocation="http://www.springframework.org/schema/beans
    http://www.springframework.org/schema/beans/spring-beans-4.1.xsd">
    <bean class="com.ssmbook2020.Customer" id="customer" p:id="300" p:name=
"孙悟空" p:male="false" p:birthday-ref="birthday"/>
    <bean class="java.util.Date" id="birthday" />
</beans>
```

（2）再次运行上面的测试类，运行结果如下。

```
Customer{id=300, name='孙悟空', male=false, birthday=Mon Jan 18 17:29:10 CST
2021}
```

经过修改以后，代码量减少了，但执行效果相同。

2.4 基于注解方式的 IoC

视频讲解

2.4.1 使用注解

在学习基于注解的 IoC 配置前，首先得有一个认知，即注解配置和 XML 配置要实现的目的是一样的，都是降低模块之间的耦合度，只是配置的形式不一样。至于实际开发中是使用 XML 还是注解，每家公司有不同的习惯，所以这两种配置方式都需要掌握。

Spring 在 IoC 部分提供了很多注解。Spring IoC 中的常用注解如表 2-5 所示。

表 2-5 IoC 容器常用注解

常用注解		说明
创建对象的注解	@Component	放在普通类上
	@Controller	放在控制器上
	@Service	放在业务对象上
	@Repository	放在持久层上
依赖注入的注解	@Autowired	1. 按类型匹配的方式注入 2. 如果有多个相同的类型，则按名字注入 3. 如果找不到名字，则抛出异常
	@Qualifier	直接按名字匹配的方式注入

Spring IoC 容器中的 XML 配置与注解可以混合使用，本书案例采用如下原则。

（1）自己编写的类通常使用注解的配置方式。

（2）引用第三方厂商开发的类使用 XML 的配置方式。

2.4.2 扫描基包

以注解的方式配置 IoC，Spring 怎样得知哪些类要放在容器中呢？这里先要配置如下的标签。

```
<context:component-scan base-package="com.ssmbook2020"/>
```

配置完该标签后，Spring 就会自动扫描 base-package 对应的包及其下面的 Java 文件，如果扫描到类上面带有@Service、@Component、@Repository、@Controller 等注解的类，则实例化这些对象并添加到 Spring 容器中。

要使用上面的标签，需要引入 context 的命名空间。

```
<beans xmlns="http://www.springframework.org/schema/beans"
    xmlns:xsi="http://www.w3.org/2001/XMLSchema-instance"
    xmlns:context="http://www.springframework.org/schema/context"
    xsi:schemaLocation="http://www.springframework.org/schema/beans
```

```
    http://www.springframework.org/schema/beans/spring-beans-4.1.xsd
    http://www.springframework.org/schema/context
    https://www.springframework.org/schema/context/spring-context-4.1.xsd">
```

下面通过具体案例来学习扫描基包的方法。

（1）创建新的 Java 工程，JDK 版本选择 1.8，工程结构如图 2-15 所示。

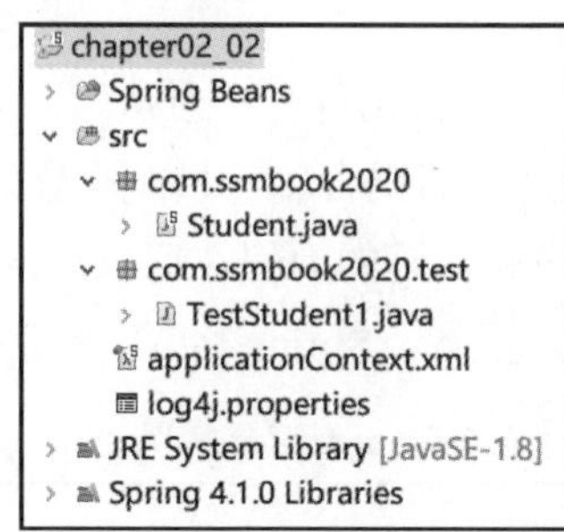

图 2-15　工程结构

（2）选择 Project → Configure Facets → Install Spring Facet 命令，添加 Spring 框架。修改 applicationContext.xml 配置文件，添加 context 的命名空间和包的扫描配置。

```
<?xml version="1.0" encoding="UTF-8"?>
<beans xmlns="http://www.springframework.org/schema/beans"
    xmlns:xsi="http://www.w3.org/2001/XMLSchema-instance"
    xmlns:context="http://www.springframework.org/schema/context"
    xsi:schemaLocation="http://www.springframework.org/schema/beans
    http://www.springframework.org/schema/beans/spring-beans-4.1.xsd
    http://www.springframework.org/schema/context
    https://www.springframework.org/schema/context/spring-context-4.1.xsd">
    <!-- 添加扫描的基包 -->
    <context:component-scan base-package="com.ssmbook2020"/>
</beans>
```

（3）在 com.ssmbook2020 包下创建学生类 Student。在类上添加@Component 注解，注解后面的字符串为放入 IoC 容器中的名字。该名字可以省略，若省略则默认为类的名字，将其首字母小写后放到容器中。

```
package com.ssmbook2020;
import java.util.Date;
import org.springframework.stereotype.Component;
/**
 * 学生类
 */
@Component("student")          //将对象放到容器中
public class Student {
    private Integer id;        //编号
    private String name;       //姓名
    private Boolean sex;       //性别：男 true，女 false
```

```
        private Date birthday;   //生日
        @Override
        public String toString(){
            return "Student [id=" + id + ", name=" + name + ", sex=" + sex + ", birthday=" + birthday + "]";
        }
    }
```

（4）编写测试类，从容器中获取名为 student 的对象，并输出对象。

```
    package com.ssmbook2020.test;
    import org.springframework.context.support.ClassPathXmlApplicationContext;
    import com.ssmbook2020.Student;

    public class TestStudent1 {
        public static void main(String[] args){
            ClassPathXmlApplicationContext context = new ClassPathXmlApplicationContext("applicationContext.xml");
            //从容器中获取学生对象，这里按类型匹配的方式从容器中获取对象，当然也可以使用以前的方法按名字从容器中获取
            Student student = context.getBean(Student.class);
            System.out.println(student);
            //关闭容器
            context.close();
        }
    }
```

（5）运行结果如下。虽然对象的属性没有值，但可以从容器中获取对象，这说明对象已经放到 IoC 容器中了。

```
    Student [id=null, name=null, sex=null, birthday=null]
```

2.4.3 IoC 容器中的注解

IoC 容器中的注解主要有以下几种。

（1）@Component 注解。

作用：把对象交给 Spring IoC 容器来管理，相当于在 XML 中配置一个 Bean。

属性：value 指定 Bean 的 id。如果不指定 value 属性，默认 Bean 的 id 是当前类的类名，首字母小写。

（2）@Controller：用于控制器的注解，在后面的 Spring MVC 框架中会用到。

（3）@Service：用于业务层 Service 的注解。

（4）@Repository：用于持久层 DAO 的注解。

后三个注解都是针对@Component 的衍生注解，它们的作用和属性是一样的。如果注解中有且只有一个属性要赋值，而且名称是 value，则 value 在赋值时可以不写。

视频讲解

2.5 依赖关系的注解

对象放在 Spring 容器中管理，那么对象之间的依赖关系是如何实现的呢？当然也是通过注解来实现的，常用的依赖关系的注解有@Autowired 、@Qualifier 和@Value，下面分别进行介绍。

2.5.1 按类型匹配注入

@Autowired 注解用于成员变量和成员方法上，自动按照类型匹配的方式注入。当该注解用在属性上时，set 方法可以省略，而且只能注入其他引用类型。它按照以下的顺序在容器中查找匹配的引用类型并进行注入。

（1）按类型匹配的方式从容器中查找对应的值注入。

（2）如果有多个匹配的类型，则按名字匹配的方式注入。

（3）如果找不到匹配的名字，则抛出异常。

按类型匹配注入的规则如图 2-16 所示。

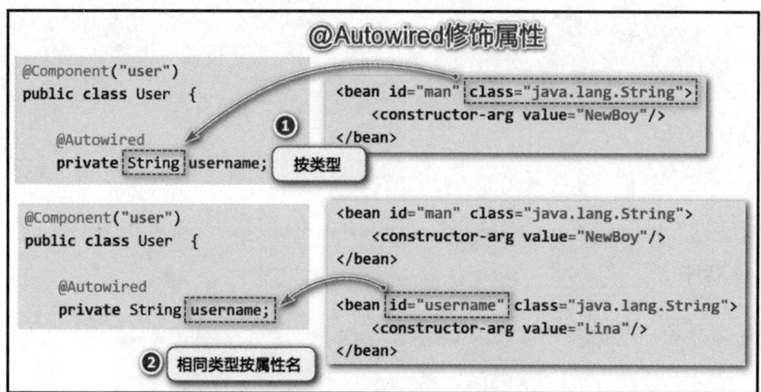

图 2-16　按类型匹配注入的规则

下面通过具体案例来学习@Autowired 注解。

（1）在 applicationContext.xml 文件中配置一个字符串对象，相当于 new String("NewBoy")，用于属性注入。

```
<bean id="man" class="java.lang.String">
  <constructor-arg value="NewBoy"/>
</bean>
```

（2）在 Student 对象的 name 属性上添加@Autowired 注解。

```
/**
 * 学生类
 */
@Component("student")          //将对象放到容器中
```

```
public class Student {
    private Integer id;       //编号
    @Autowired
    private String name;      //姓名
    private Boolean sex;      //性别：男 true，女 false
    private Date birthday;    //生日
}
```

（3）运行测试类，可以发现 name 属性已被赋值。

```
Student [id=null, name=NewBoy, sex=null, birthday=null]
```

（4）修改 applicationContext.xml 配置文件，再添加一个 String 的对象，这时容器中有两个 String 类型，但第二个的名字叫 name，该对象会按名字匹配的方式注入属性值。

```
<bean id="man" class="java.lang.String">
  <constructor-arg value="NewBoy"/>
</bean>
<bean id="name" class="java.lang.String">
  <constructor-arg value="Lina"/>
</bean>
```

（5）再次运行测试类，会发现属性值变成了第二个字符串的值。

```
Student [id=null, name=Lina, sex=null, birthday=null]
```

（6）修改第二个 String 的 id 为 woman，这时容器中就找不到 id 为 name 的名字了，运行测试类会抛出异常。

```
<bean id="man" class="java.lang.String">
  <constructor-arg value="NewBoy"/>
</bean>
<bean id="woman" class="java.lang.String">
  <constructor-arg value="Lina"/>
</bean>
```

（7）错误信息：期望是一个匹配的值，但找到了 man 和 woman 两个值。

```
org.springframework.beans.factory.NoUniqueBeanDefinitionException:    No
qualifying bean of type [java.lang.String] is defined: expected single matching
bean but found 2: man,woman
```

通过以上的案例可以知道@Autowired 的注入规则，该注解也可以用在成员方法上，注入规则也是一样的，这是一个常用的注解。

2.5.2　按名字匹配注入

@Qualifier 注解有以下几个特点。

（1）必须与@Autowired 配合使用，不能单独使用。

（2）可以放在成员变量或成员方法上。

（3）按名字匹配的方式注入。

（4）value 属性指定要注入的名字。

按名字匹配注入的规则如图 2-17 所示。

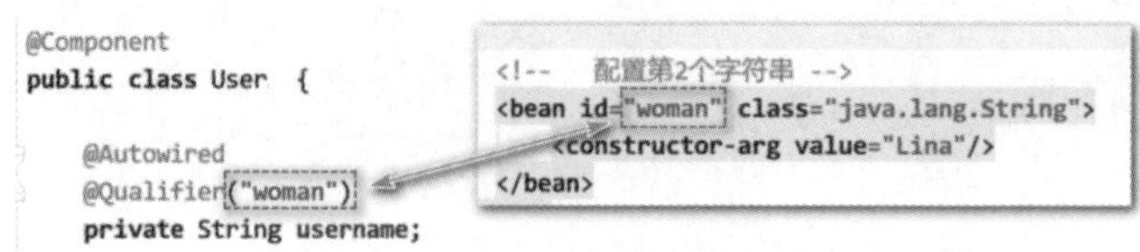

图 2-17　按名字匹配注入的规则

下面通过具体案例来学习@Qualifier 注解。在上个案例的基础上修改 Student 类，给 name 属性再添加一个@Qualifier 注解，指定注解的名字为 woman。

```
@Component("student")         //将对象放到容器中
public class Student {
   private Integer id;        //编号

   @Autowired
   @Qualifier("woman")
   private String name;       //姓名
   private Boolean sex;       //性别：男 true，女 false
   private Date birthday;     //生日
   @Override
   public String toString(){
      return "Student [id=" + id + ", name=" + name + ", sex=" + sex + ", birthday=" + birthday + "]";
   }
}
```

运行测试类，发现不再报错，成功注入了属性值，这是因为指定了按名字匹配的方式注入。

```
Student [id=null, name=Lina, sex=null, birthday=null]
```

如果在容器中要优先按名字匹配，就同时使用@Autowired 和@Qualifier 注解，以后可以根据实际情况使用。

2.5.3　注入简单类型

1. @Value 注解

@Autowired 注解是用于引用类型的对象注入，对于一些简单类型的注入可以使用@Value 注解。@Value 也可以注入日期类型。

@Value 有@Value("${}")和@Value("#{}")两种取值方式。第 1 种方式主要从 Java 的属性文件中读取值，第 2 种方式则是通过 SpEL 表达式读取相应的值。

2. 直接注入

（1）修改 Student 类，给生日注入一个日期，按 yyyy/mm/dd 的格式编写。

```
@Value("2021/02/25")
private Date birthday;  //生日
```

（2）运行测试类，结果如下。

```
Student [id=null, name=Lina, sex=null, birthday=Thu Feb 25 00:00:00 CST 2021]
```

由上述结果可以发现生日被直接注入了值，这种场景在实际开发中使用不多，后期主要是读取属性文件注入，具体案例如下。

3. 读取属性文件注入

下面通过读取属性文件 student.properties 中的值来注入 Student 对象中。

（1）在 src 目录下创建属性文件 student.properties，其内容如下。

```
student.id=100
student.sex=true
```

（2）在 applicationContext.xml 文件中读取属性配置文件。

```
<!-- 读取属性配置文件 -->
<context:property-placeholder location="classpath:student.properties"/>
```

（3）在 Student 类的 id 和 sex 属性上使用@Value 注解，注入属性值。

```
@Value("${student.id}")
private Integer id;
@Value("${student.sex}")
private Boolean sex;
```

（4）运行测试类，结果如下。

```
Student [id=100, name=Lina, sex=true, birthday=Thu Feb 25 00:00:00 CST 2021]
```

至此，每个属性都注入了，但使用的是不同的方式，完整代码如下。

（1）Student.java 主控文件。

```
package com.ssmbook2020;
import java.util.Date;
/**
 * 学生类
 */
@Component("student")         //将对象放到容器中
public class Student {
    @Value("${student.id}")
    private Integer id;       //编号
    @Autowired
    @Qualifier("woman")
    private String name;      //姓名
```

```
    @Value("${student.sex}")
    private Boolean sex;     //性别：男 true，女 false
    @Value("2021/02/25")
    private Date birthday;   //生日
}
```

（2）applicationContext.xml 配置文件。

```
<?xml version="1.0" encoding="UTF-8"?>
<beans xmlns="http://www.springframework.org/schema/beans"
    xmlns:xsi="http://www.w3.org/2001/XMLSchema-instance"
    xmlns:context="http://www.springframework.org/schema/context"
    xsi:schemaLocation="http://www.springframework.org/schema/beans
    http://www.springframework.org/schema/beans/spring-beans-4.1.xsd
    http://www.springframework.org/schema/context
    https://www.springframework.org/schema/context/spring-context-4.1.xsd">
    <!-- 添加扫描的基包  -->
    <context:component-scan base-package="com.ssmbook2020"/>
    <!-- 读取属性配置文件 -->
    <context:property-placeholder location= "classpath:student.properties"/>
    <bean id="man" class="java.lang.String">
       <constructor-arg value="NewBoy"/>
    </bean>
    <bean id="woman" class="java.lang.String">
       <constructor-arg value="Lina"/>
    </bean>
</beans>
```

2.6 本章小结

本章对 Spring 框架进行了简要介绍，重点讲解了 IoC 和 AOP 的概念。通过一个入门案例，介绍了 IoC 容器的使用方法，在 Spring 中有多种创建 Spring 容器的方式；详细介绍了 Bean 标签的属性，并且使用案例一一说明。

IoC 容器中一个比较重要的功能就是依赖注入，本章介绍了 3 种依赖注入的方式：构造器注入、set 方式注入和 p 命名空间注入。

此外，本章还详细介绍了 Spring 中支持的注解。

习题 2

动手完成本章介绍的全部案例。

第3章 Spring AOP

本章学习内容

- Spring AOP 概述；
- 动态代理模式；
- AspectJ 表达式；
- 使用 XML 配置实现 AOP；
- 使用注解实现 AOP。

3.1 Spring AOP 概述

3.1.1 AOP 的概念

AOP 是 Spring 框架中最核心的一个功能，首先通过一个生活中的案例来学习一下什么是 AOP。比如银行系统会有一个取款流程，传统程序的流程如图 3-1 所示。

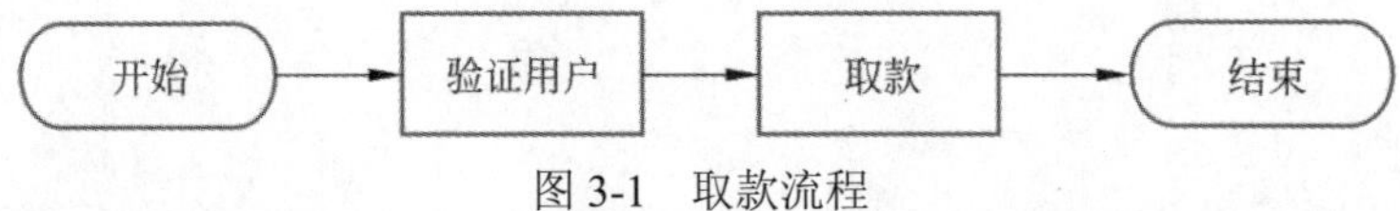

图 3-1　取款流程

假设系统还有一个查询并显示余额的流程，如图 3-2 所示。

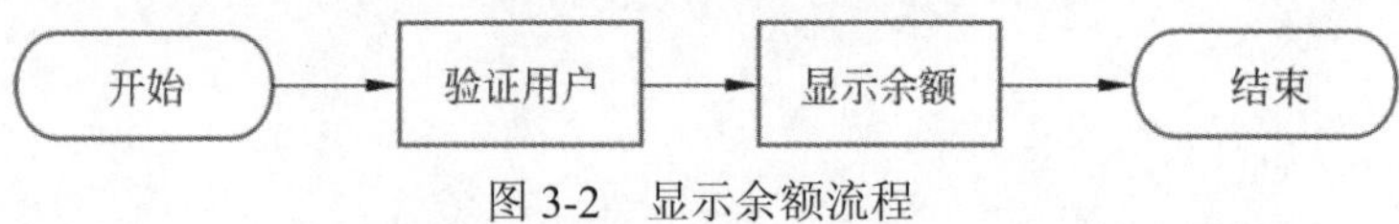

图 3-2　显示余额流程

把这两个流程放到一起，会发现两者有一个相同的验证流程，如图 3-3 所示。

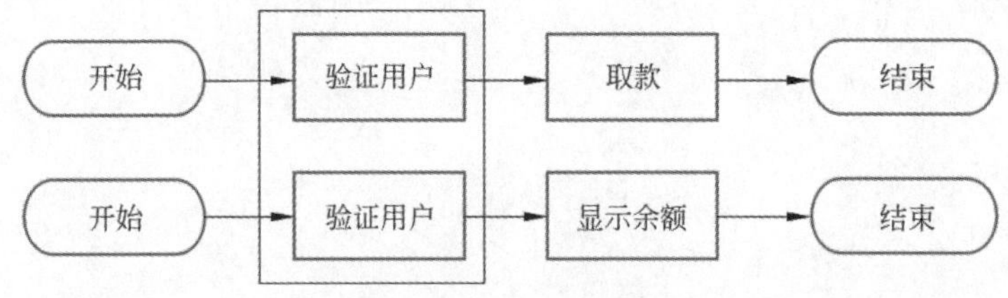

图 3-3　验证流程

在系统开发中，验证用户的功能是相同的，但又在不同的地方出现，可以把这部分代码提取出来写成一个共用的类，这个类就称为切面（Aspect）类。切面类中的方法（验证用户）称为通知（Advice），在程序执行时动态地添加到主业务类的方法（取款和显示余额）前后，这个主业务类的方法称为连接点（Joinpoint）。验证用户的切面类可以在不同的地方重用，这种编程方式叫作 AOP（Aspect Oriented Programming），即面向切面编程。

有了 AOP，在编写代码时就不需要把验证用户的步骤写进去，可以完全不考虑验证用户的功能，只编写取款和显示余额的业务代码，而在另一个地方写好验证用户的代码。这个验证用户的代码就是切面代码，以后在执行取款和显示余额时，将验证用户的功能在执行取款和显示余额前调用。

代码在 Spring 容器中执行时，通过配置告诉 Spring 这段代码要添加的地方，Spring 就会在执行正常业务流程时把验证代码和取款代码织入在一起。

AOP 的真正目的是让程序员在编写代码时只需考虑主要业务流程，而不用考虑和业务无关却又必须要写的代码。

3.1.2 AOP 中类与切面的关系

AOP 的本质是在一系列纵向的控制流程中，把那些相同的子流程（如验证用户）提取成一个横向的面，将分散在主流程中相同的代码提取出来，然后在程序编译或运行时，将这些提取出来的切面代码应用到需要执行的地方。

例如取款、查询、转账前都要进行用户验证，则验证用户就可以做成切面类。在执行取款、查询、转账的操作时，由 Spring 容器将验证用户的代码织入在它们的前面，从而达到验证用户的目的。验证用户的代码只需要编写一次，可以让程序员将编程的精力放在取款、查询、转账等主要业务上。切面与主业务类的关系如图 3-4 所示。

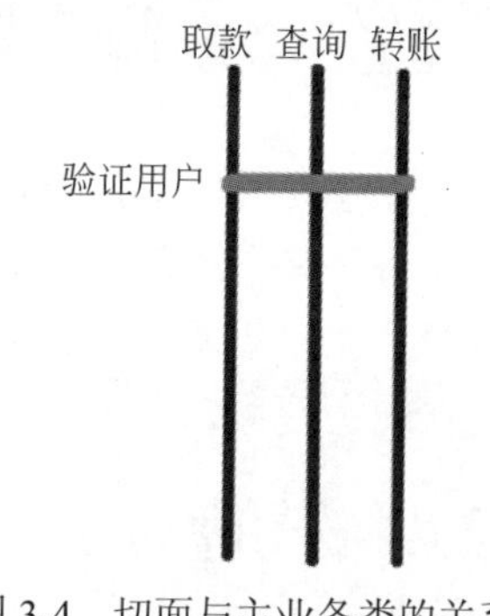

图 3-4　切面与主业务类的关系

3.1.3 AOP 的应用场景

面向切面编程的主要场景有如下几种，切面可以分别在类 1 和类 2 方法中加入事务、日志、权限控制等功能，如图 3-5 所示。上面的验证用户就是权限控制的一种。

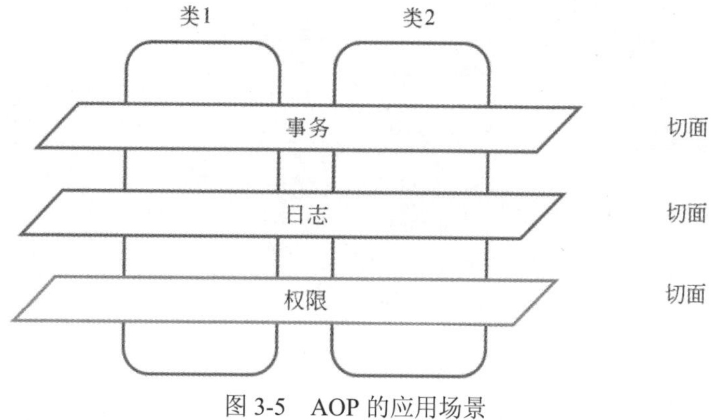

图 3-5　AOP 的应用场景

比如日志记录功能，服务器端重要的操作步骤是需要用日志记录下来，便于以后服务器的管理和维护，所以系统中就会出现类似如下代码。

```
logger.info("管理员登录");              //日志记录
userService.login();                    //业务操作
logger.info("管理员删除用户");          //日志记录
userService.deleteUser();               //业务操作
logger.info("管理员退出");              //日志记录
userService.logout();                   //业务操作
```

上面的业务代码和日志记录代码会分布在整个系统中，而且是零散的。

几乎所有的重要操作方法前面都会加上日志记录代码，这样的代码写起来烦琐，不但占用了开发时间和精力，而且不容易维护。因此可以将日志记录的功能做成切面类，让程序在执行时再动态地将日志与主业务功能织入在一起。

AOP 的核心实现是动态代理模式，下面介绍一下 Java 中常用的 JDK 动态代理。

3.2 动态代理模式

代理模式的作用是为其他对象提供一种代理以便控制该对象的访问。代理模式可以详细控制访问某个对象的方法，在调用这个方法前做一些前置处理，调用这个方法后也可以做后置处理。代理模式的实现分为静态代理和动态代理，在 Spring 中多使用动态代理。动态代理的实现可以分为 JDK 动态代理和 CGLIB 动态代理，下文重点介绍 JDK 动态代理。

3.2.1 代理模式对象

JDK 动态代理等代理模式所涉及的对象如图 3-6 所示，如租客、中介、房东等都是代理对象。

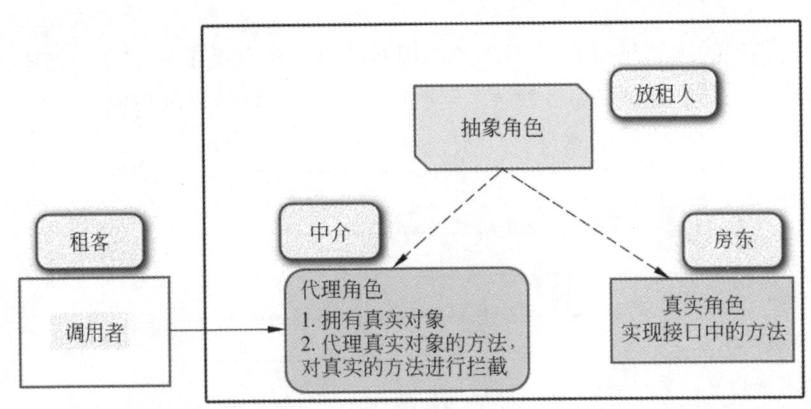

图 3-6　代理模式示意图

在代理模式中有以下 4 种角色。

（1）真实角色：需要实现抽象角色的接口，定义了真实角色所要实现的业务逻辑，即真正的业务逻辑。

（2）代理角色：相当于真实角色的一个代理角色，可以改写真实角色的方法或对真实角色的方法进行拦截，并可以附加自己的操作。

（3）抽象角色：指代理角色和真实角色对外提供的公共方法，定义了真实角色的行为，一般为一个抽象类或接口。

（4）调用者：使用真实角色的调用者，不属于代理模式中的一部分。

3.2.2 JDK 动态代理

JDK 动态代理必须借助于一个接口才能产生代理对象。因此，对于使用业务接口的类，Spring 默认使用 JDK 动态代理实现 AOP。

1. 动态代理的特点

（1）程序在执行过程中动态生成代理对象，不用手动编写代理对象。

（2）不需要重写目标对象中所有同名的方法，只关注需要代理的方法即可。

2. 动态代理类相应的 API

1）Proxy 类

```
    static Object newProxyInstance(ClassLoader loader, Class[] interfaces, InvocationHandler h)
```

在 JDK 的 API 中存在一个 Proxy 类，其中有一个生成动态代理对象的方法 newProxyInstance()。

该类的参数说明如下。

（1）loader：真实对象的类加载器。

（2）interfaces：真实对象所有实现的接口数组。

（3）h：具体的代理操作，InvocationHandler 是一个接口，需要传入一个实现了此接口的实现类。

（4）返回值：生成的代理对象。

2）InvocationHandler 接口

```
    Object invoke(Object proxy, Method method, Object[] args)
```

在这个方法中实现对真实方法的增强或拦截，其参数说明如下。

（1）proxy：即 newProxyInstance()方法返回的代理对象，该对象一般不在 invoke 方法中使用，容易出现递归调用。

（2）method：真实对象的方法对象，会进行多次调用，每次调用 method 对象都不同。

（3）args：代理对象调用方法时传递的参数。

（4）返回值：真实对象方法的返回值。

下面通过一个案例演示如何使用 JDK 动态代理实现 Spring AOP。

假设有一个出租的接口，其代码如下。

```
package com.ssmbook2020;
/**
 * 出租的接口
 */
public interface Lease {
    /**
     * 定义出租的行为
     * @param money 租金
```

```
     */
    void rentOut(int money);
}
```

真实角色是房东，其代码如下。

```
/**
 * 房东：真实角色
 */
public class Landlord implements Lease {
    @Override
    public void rentOut(int money){
        System.out.println("房东出租房子，收取租金：" + money);
    }
}
```

该房屋出租案例的开发流程如下。

（1）直接创建真实对象，并调用真实对象的方法。

（2）通过 Proxy 类创建代理对象，并调用代理对象的方法。

（3）分别输出真实对象和代理对象的实现类。

```
package com.ssmbook2020;
import java.lang.reflect.InvocationHandler;
import java.lang.reflect.Method;
import java.lang.reflect.Proxy;
/**
 * 租客：使用代理对象的调用者
 */
public class Tenant {
    public static void main(String[] args){
        //(1)直接找房东租房(创建真实对象)
        Lease landlord = new Landlord();
        //调用真实对象的方法
        landlord.rentOut(2000);
        System.out.println("真实对象：" + landlord.getClass());
        //输出一条线分隔开
        System.out.println("==========");
        //(2)找中介租房 (创建代理对象)
        Lease agent = (Lease)
        Proxy.newProxyInstance(landlord.getClass(). getClassLoader();
 //真实对象的类加载器
               landlord.getClass().getInterfaces();
//获取真实对象所有实现的接口
               new InvocationHandler() { //代理的实现
                   /**
                    * proxy：即 newProxyInstance()方法返回的代理对象
                    * method：真实对象的方法对象，会被调用多次，每次调用 method 对象
                    * 都不同
```

```
                    * args：代理对象调用方法时传递的参数
                    */
                   @Override
        public Object invoke(Object proxy, Method method, Object[] args)
throws Throwable{
                        //如果是出租的方法
                        if (method.getName().equals("rentOut")){
                           //对真实方法进行代理，但不修改原来类的方法
   System.out.println("中介出租房子，收取中介费：200");
                        }
                        //调用真实对象的方法
                        return method.invoke(landlord, args);
                     }
                  });
        //调用代理对象的方法
        agent.rentOut(2000);
        //输出代理对象
        System.out.println("代理对象：" + agent.getClass());
     }

   }
```

在以上代码中，代理角色是程序在执行过程中动态生成的，在没有修改真实对象的前提下对真实对象进行了代理，并添加了新的功能。

该案例的代码执行效果如下。

```
房东出租房子，收取租金：2000
真实对象：class com.ssmbook2020.Landlord
==========
中介出租房子，收取中介费：200
房东出租房子，收取租金：2000
代理对象：class com.sun.proxy.$Proxy0
```

可以看到，代理对象实现类的名字是 com.sun.proxy.$Proxy0，在后面的 Spring AOP 中还会再用到。

3.3 AOP 的实现

3.3.1 AOP 的常用增强类型

视频讲解

AOP 的实现原理其实就是使用代理模式，由 AOP 框架动态生成一个代理对象，该对象可作为代替目标对象使用。Spring 中的 AOP 代理由 Spring 的 IoC 容器负责生成、管理，因此 AOP 代理可以直接使用容器中的其他 Bean 实例作为目标，这种关系可由 IoC 容器的依赖注入提供。

常用的 AOP 代理增强主要包括前置增强、后置增强、环绕增强、异常增强等。

（1）前置增强（Before Advice）：在某连接点之前执行的增强，但这个增强不能阻止连接点之前的执行流程（除非它抛出一个异常）。

（2）后置增强（After Returning Advice）：在某连接点正常完成后执行的增强。例如，一个方法没有抛出任何异常，正常返回。

（3）异常增强（After Throwing Advice）：在方法抛出异常退出时执行的增强。

（4）最终增强（After (Finally) Advice）：当某连接点退出时执行的增强（无论是正常返回还是异常退出）。

（5）环绕增强（Around Advice）：包围一个连接点的增强，如方法调用。这是最强大的一种增强类型。环绕增强可以在方法调用前后完成自定义的行为，它会选择是否继续执行连接点、直接返回它自己的返回值或抛出异常来结束执行。

在内部调用这 5 种增强类型的组织方式如下。

```
try {
调用前置增强
环绕前置处理
调用目标对象方法
环绕后置处理
调用后置增强
} catch(Exception e) {
调用异常增强
} finally {
调用最终增强
}
```

视频讲解

3.3.2 AspectJ 表达式

AspectJ 表达式又称为切入点表达式，它其实是一组规则，指定哪些包下的哪些类和方法织入通知代码。

1. 切点函数

常用的切点函数有以下 3 个，如表 3-1 所示。

表 3-1 切点函数

切点函数	作　用
execution	细粒度函数，可以精确到类中的某个方法
within	粗粒度函数，只能精确到类
bean	粗粒度函数，只能精确到类，它是通过 id 从容器中获取对象

2. 表达式语法

图 3-7 是 Spring 官方文档对表达式语法的介绍，其中？号表示出现 0 次或 1 次。

切点函数中共有 6 个参数可以指定，只有“返回类型”、“方法名”和“参数列表”是必须指定的，参数中可以使用通配符，不同位置的通配符的含义有所区别。

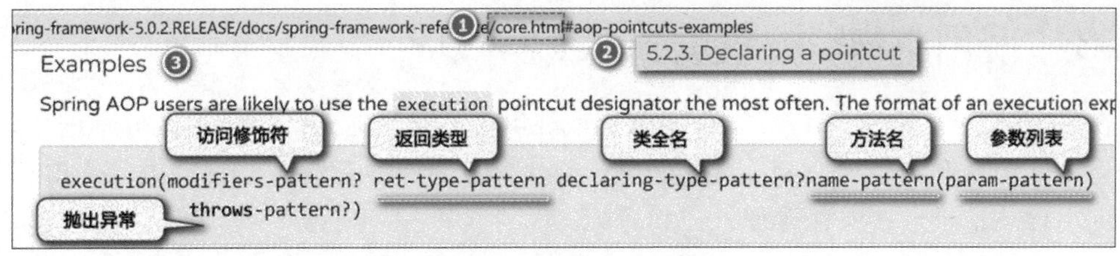

图 3-7　切点函数的语法

- 方法中参数个数通配符写法：

() 表示没有参数。

(*) 表示 1 个任意类型的参数。

(..) 表示 0 个或多个参数。

- 类全名的包通配符写法：

.. 表示当前包和子包。

3. 举例说明

（1）最精确的写法。

```
execution(public void com.ssmbook2020.service.impl.AccountServiceImpl.
save())
```

（2）匹配 Service 的包和子包下面所有的类和方法，方法参数是 String 类型。

```
execution(* com.ssmbook2020.service..*.*(String))
```

（3）覆盖最全的写法，匹配所有的类和方法。

```
execution(* *(..))
```

（4）匹配方法名是 save 或 update 的方法。匹配时也可以使用&&符号，虽然语法是正确的，但没有意义。

```
execution(* save(..)) || execution(* update(..))
```

（5）除了方法名是 save 的所有方法。

```
!execution(* save(..))
```

（6）匹配包和子包中的所有类，只精确到类。

```
within(com.itheima..*)
```

（7）从容器中获取一个 id 为 accountService 的类中的所有方法，只精确到类。

```
bean(accountService)
```

（8）从容器中获取所有 Service 的方法，只精确到类。

```
bean(*Service)
```

3.3.3 使用 XML 配置方式实现 AOP

在 Spring 中，AOP 编程可以使用配置或注解两种实现方式，下面对此分别进行介绍。首先通过一个案例来介绍 XML 配置方式，该案例实现的功能是：当向数据库中保存账户时，使用日志记录这次保存操作。

1. 开发流程

该案例的开发步骤如下。

（1）开发业务类，主业务方法为添加账户。

（2）开发切面类，用于在每个主业务方法执行前添加记录日志的功能。

（3）使用 AOP 将业务类与切面类织入在一起，实现需求的功能。

2. 案例结构

整个案例的工程结构如图 3-8 所示。

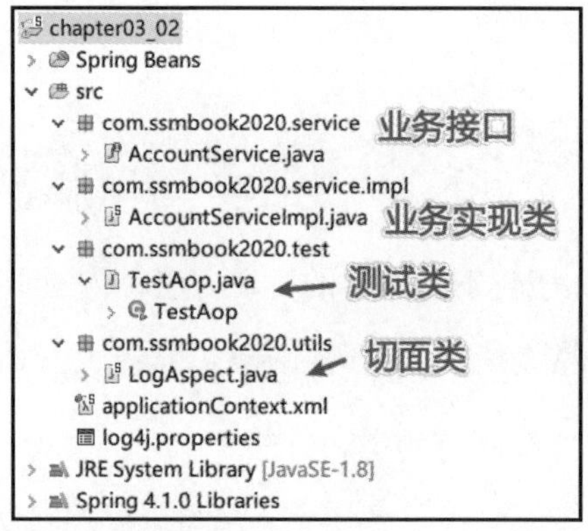

图 3-8　XML 配置方式的案例结构

3. 业务接口和实现类

（1）建立账户业务接口并保存账户。

```
package com.ssmbook2020.service;
/**
 * 账户业务接口
 */
public interface AccountService {
    /**
     * 保存账户
     */
    void save();
}
```

（2）实现业务接口类，输出“保存账户”。

```
package com.ssmbook2020.service.impl;
import com.ssmbook2020.service.AccountService;
```

```
/**
 * 实现类
 */
public class AccountServiceImpl implements AccountService {
    @Override
    public void save() {
        System.out.println("保存账户");
    }
}
```

4. 记录日志的工具类

```
package com.ssmbook2020.utils;
import java.sql.Timestamp;
/**
 * 记录日志功能的类：切面类 = 切入点(规则)+通知(方法)
 */
public class LogAspect {
    /**
     * 记录日志
     */
    public void printLog() {
        System.out.println(new Timestamp(System.currentTimeMillis()) + " 记
录日志");
    }
}
```

5. AOP 的配置

1）AOP 配置结构图

图 3-9 为 Spring AOP 的配置结构图。在一个项目中只需配置一次。读者在编写 XML 配置时可以参照图 3-9 进行编写，避免出错。

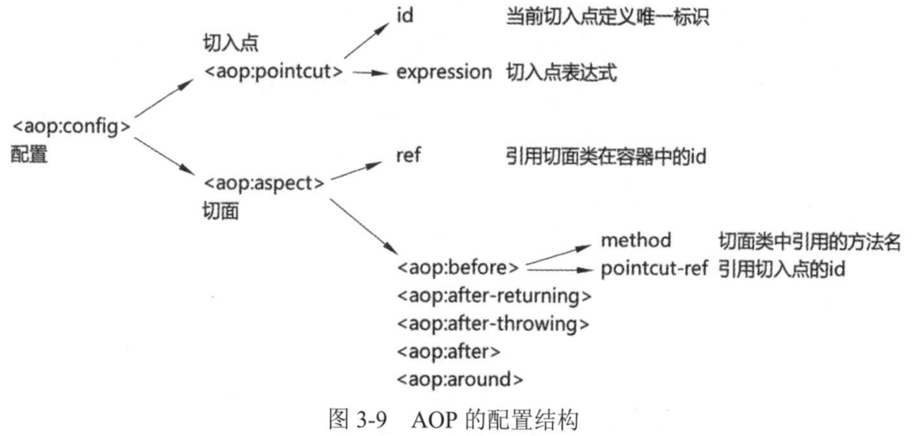

图 3-9 AOP 的配置结构

2）配置文件

接下来在 xml 配置文件中编写 AOP 的配置，主要步骤如下。

（1）配置日志记录类，包括切面类 LogAspect。

（2）配置正常的业务类 AccountServiceImpl。

（3）进行 AOP 配置，配置流程参考图 3-9。需要注意的是，XML 的 Schema 需要导入 AOP 的命名空间。

```
<?xml version="1.0" encoding="UTF-8"?>
<beans
    xmlns="http://www.springframework.org/schema/beans"
    xmlns:xsi="http://www.w3.org/2001/XMLSchema-instance"
    xmlns:p="http://www.springframework.org/schema/p"
    xmlns:aop="http://www.springframework.org/schema/aop"
    xsi:schemaLocation="http://www.springframework.org/schema/beans
    http://www.springframework.org/schema/beans/spring-beans-4.1.xsd
    http://www.springframework.org/schema/aop
    https://www.springframework.org/schema/aop/spring-aop-4.1.xsd">
     <!--正常业务对象-->
    <bean class="com.ssmbook2020.service.impl.AccountServiceImpl" id=
"accountService"/>
    <!-- 切面类：日志记录对象 -->
    <bean class="com.ssmbook2020.utils.LogAspect" id="logAspect"/>

    <!-- 编写 Aop 的配置，要导入 AOP 的命名空间 -->
    <aop:config>
        <!--
            配置切入点，通过切入点表达式来配置
             id：给表达式定义唯一标识
             expression：使用切入点函数定义表达式，语法为“访问修饰符 返回类型 包名.
类名.方法名(参数类型) 抛出异常类型” -->
        <aop:pointcut id="pt" expression="execution(public void
com.ssmbook2020.service.impl.AccountServiceImpl.save())"/>

        <!-- 切面配置， ref 引用切面对象 id -->
        <aop:aspect ref="logAspect">
            <!-- 使用什么类型的通知：前置通知，后置通知等
            method：表示切面中的方法名字 pointcut-ref：引用上面的切入点表达式 -->
            <aop:before method="printLog" pointcut-ref="pt"/>
        </aop:aspect>
    </aop:config>
</beans>
```

3）测试类

配置完成 AOP 后，开始编写测试类，其步骤如下。

（1）调用业务方法。

（2）输出业务类的 getClass()，查看输出的代理类对象。

```
package com.ssmbook2020.test;
```

```
    import org.springframework.context.support.ClassPathXmlApplicationContext;
    import com.ssmbook2020.service.AccountService;
    public class TestAop {
        public static void main(String[] args){
            ClassPathXmlApplicationContext context = new
ClassPathXmlApplicationContext("applicationContext.xml");
            //从容器中获取对象
            AccountService accountService = context.getBean(AccountService.class);
            //得到的是代理对象
            System.out.println(accountService.getClass());
            //调用业务方法
            accountService.save();
            //关闭容器
            context.close();
        }
    }
```

测试类的运行结果如下。

```
class com.sun.proxy.$Proxy5
2021-01-20 12:22:11.373 记录日志
保存账户
```

AOP 的配置本质上是使用代理模式实现功能增强，不需要自己编写代理模式，而通过配置就可以实现。由此可以看到，此时的 AccountService 的实现类其实是个代理对象。

4）执行流程分析

因为配置了 AOP，分开编写的切面类 LogAspect 和正常业务类 AccountServiceImpl 被 Spring AOP 框架在执行时织入在一起，生成了一个代理对象，如图 3-10 所示。

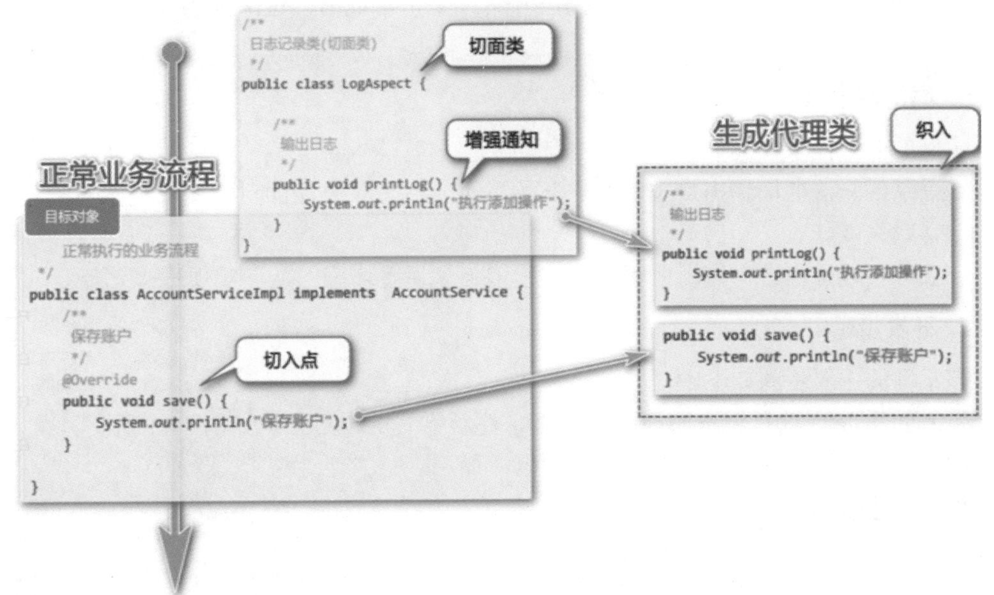

图 3-10　织入的流程

Spring AOP 其实使用了两种动态代理实现方式，如果一个类并没有实现任何的接口，则无法使用上面所说的 JDK 动态代理，这时需要使用 CGLIB 代理。CGLIB 代理的本质是对原有类的继承，子类重写相应的方法，其生成过程与 JDK 类似，这里不再赘述。

视频讲解

3.3.4 使用注解方式实现 AOP

通过 3.3.3 节的案例可以知道，使用 XML 的方式配置 AOP 是比较烦琐的，如何简化 XML 的配置呢？可以使用注解的方式。

对 Spring AOP 中常用的 Aspect 注解进行如下介绍。

（1）@Aspect：放在类的上面，表示这是一个切面类。

（2）@Before：前置增强。它有以下两个参数。

- value：该成员用于定义切点。
- execution：切点函数，告诉 Spring 在哪些地方进行前置增强的织入。

（3）@AfterReturning：后置增强。它有以下 4 个参数。

- value: 该成员用于定义切点。
- pointcut：表示切点的信息。如果指定 pointcut 值，将覆盖 value 的值，可以理解为它们的作用是相同的。
- returning：将目标对象方法的返回值绑定给增强的方法，返回值的名字要与实际返回的变量名相同。
- argNames：同 returning。

（4）@Around：环绕增强。它有 value 和 argNames 两个参数，其含义同上。

（5）@AfterThrowing：异常增强，有以下 4 个参数。

- value、pointcut、argNames 同上。
- throwing：将抛出的异常绑定到增强方法中。

（6）@After：最终增强。不管是抛出异常或是正常退出，该增强都会得到执行，它有 Value 和 arg Names 两个参数，其含义同上。

1. 案例需求

案例的业务需求描述如下。

在登录的方法前面输出日志。用户开始登录，在登录方法的后面输出提示日志：用户登录成功或失败。

在业务实现过程中，使用注解定义前置增强和后置增强，从而实现日志功能。

2. 开发步骤

该案例的主要步骤如下。

（1）创建 Java 项目，添加 Spring 框架，注解方式的项目结构如图 3-11 所示。

（2）编写各个类的代码如下，注意代码注释。

- User.java 用户实体类。

```
package com.ssmbook2020.entity;
/**
```

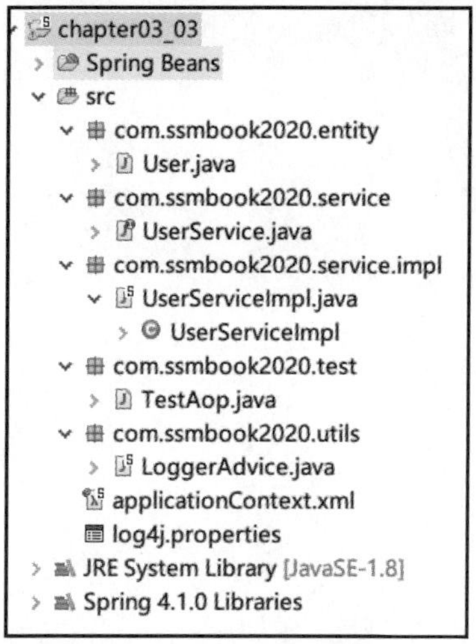

```
chapter03_03
  > Spring Beans
  v src
    v com.ssmbook2020.entity
      > User.java
    v com.ssmbook2020.service
      > UserService.java
    v com.ssmbook2020.service.impl
      v UserServiceImpl.java
        > UserServiceImpl
    v com.ssmbook2020.test
      > TestAop.java
    v com.ssmbook2020.utils
      > LoggerAdvice.java
      applicationContext.xml
      log4j.properties
  > JRE System Library [JavaSE-1.8]
  > Spring 4.1.0 Libraries
```

图 3-11　注解方式的项目结构

```
 * 用户实体类对象
 */
public class User {
    private int id;             //主键
    private String name;        //用户名
    private String password;    //密码

    public User(int id, String name, String password) { //带全参的构造方法
        super();
        this.id = id;
        this.name = name;
        this.password = password;
    }

    public User() {             //默认无参的构造方法
        super();
    }
    //省略了 get 和 set 方法
}
```

- UserService.java 业务接口类。

```
package com.ssmbook2020.service;
import com.ssmbook2020.entity.User;
/**
 * 用户的业务接口
 */
public interface UserService {
    /**
     * 登录的方法
     * @param name 用户名
```

```
     * @param password 密码
     * @return 登录成功返回User对象，登录失败返回null
     */
    public User login(String name,String password);
}
```

- UserServiceImpl.java 业务实现类。

```
package com.ssmbook2020.service.impl;
import com.ssmbook2020.entity.User;
import com.ssmbook2020.service.UserService;
/**
 * 用户业务类的实现
*/
public class UserServiceImpl implements UserService {
    @Override
    public User login(String name, String password){
        System.out.println("业务方法login运行，正在登录...");
        //登录成功
        if ("newboy".equals(name) && "520".equals(password)){
            return new User(100, "newboy", "520");
        }
        //登录失败
        return null;
    }
}
```

- LoggerAdvice.java 切面类。

```
package com.ssmbook2020.utils;
import java.sql.Timestamp;
import org.apache.log4j.Logger;
import org.aspectj.lang.JoinPoint;
import org.aspectj.lang.annotation.AfterReturning;
import org.aspectj.lang.annotation.Aspect;
import org.aspectj.lang.annotation.Before;

/** 需要织入的日志切面类 */
@Aspect
public class LoggerAdvice {
    //log4j日志类
    Logger logger = Logger.getLogger(LoggerAdvice.class);

    //后置增强
    @AfterReturning(pointcut = "execution(* com.ssmbook2020..*.*(..))",
returning = "ret")
    public void afterReturning(JoinPoint join, Object ret){
        String method = join.getSignature().getName();
```

```
        Object[] args = join.getArgs();
        if ("login".equals(method)){
            if (ret != null){
                logger.info(new Timestamp(System.currentTimeMillis()) + " "
+ args[0] + "登录成功");
            } else {
                logger.info(new Timestamp(System.currentTimeMillis()) + " "
+ args[0] + "登录失败");
            }
        }
    }

    //前置增强
    @Before(value = "execution(* com.ssmbook2020..*.*(..))")
    public void methodBefore(JoinPoint join){
        String method = join.getSignature().getName();
        Object[] args = join.getArgs();
        if ("login".equals(method)){
            logger.info(new Timestamp(System.currentTimeMillis()) + " " +
args[0] + "开始登录");
        }
    }
}
```

（3）配置 applicationContext.xml 文件，在 XML 文件头部添加 AOP 命名空间，以使用与 AOP 相关的标签。同时因为代码中用到了 p 的方式注解，所以也添加了 p 命名空间。

```
<?xml version="1.0" encoding="UTF-8"?>
<beans xmlns="http://www.springframework.org/schema/beans"
    xmlns:xsi="http://www.w3.org/2001/XMLSchema-instance"
    xmlns:p="http://www.springframework.org/schema/p"
    xmlns:aop="http://www.springframework.org/schema/aop"
    xsi:schemaLocation="http://www.springframework.org/schema/beans
http://www.springframework.org/schema/beans/spring-beans-4.1.xsd
http://www.springframework.org/schema/aop
http://www.springframework.org/schema/aop/spring-aop-4.1.xsd">
    <!-- 日志记录类 -->
    <bean id="loggerAdvice" class="com.ssmbook2020.utils.LoggerAdvice" />
    <!-- 业务类 -->
    <bean id="userService" class="com.ssmbook2020.service.impl.
UserServiceImpl" />
    <!-- 织入使用注解定义的增强，需要引入 AOP 命名空间 -->
    <aop:aspectj-autoproxy />
</beans>
```

上面的配置将所有的 JavaBean 加入 Spring 容器中，其中最重要的是<aop:aspectj-autoproxy />，表示所有的 AOP 自动代理，通过注解的方式织入。

（4）构建测试类，其代码如下。

```
package com.ssmbook2020.test;
import org.springframework.context.support.ClassPathXmlApplicationContext;
import com.ssmbook2020.service.UserService;
public class TestAop {
    public static void main(String[] args){
        ClassPathXmlApplicationContext context = new
ClassPathXmlApplicationContext("applicationContext.xml");
        //得到业务类
        UserService userService = (UserService) context.getBean("userService");
        //运行业务登录方法
        userService.login("newboy", "520");
        context.close();
    }
}
```

测试类代码的运行结果如下。

```
INFO - 2021-01-20 12:48:22.617 newboy 开始登录
业务方法 login 运行，正在登录...
INFO - 2021-01-20 12:48:22.618 newboy 登录成功
```

如果把 userService.login("newboy", "520")换成 userService.login("Lina", "1314")，则运行结果如下。

```
INFO - 2021-01-20 12:48:05.503 Lina 开始登录
业务方法 login 运行，正在登录...
INFO - 2021-01-20 12:48:05.504 Lina 登录失败
```

在业务类的代码运行时，该方法的前后各输出了日志的内容，这就是代码的织入，也称为前置或后置增强。

通过使用注解，使得 Spring AOP 的配置被极大地简化。如果把业务类和切面类在 Spring IoC 中的配置也改成注解，则可以进一步简化 XML 的配置内容。

3.4 本章小结

本章介绍了 Spring 的另一个重要特性 AOP，它是 Spring 框架最核心、最基础的技术之一。Spring AOP 的实现原理是使用了动态代理模式，在此重点介绍了 JDK 的代理模式。

本章分别讲解了 XML 配置和注解两种方式实现 Spring AOP，注解的方式相对代码量更少，后期使用比较多。对 Spring AOP 的学习只是 Spring 框架学习的开始，它是一个庞大的框架，其目标是让一切 Java EE 的开发都变得更简洁。

习 题 3

1. 什么是 Spring AOP？
2. Spring AOP 的实现原理是什么？
3. 分别使用 XML 配置和注解方式实现本章案例。

第 4 章 Spring JDBC

本章学习内容

- Spring JDBC 模块；
- JdbcTemplate 的使用；
- 数据源的配置。

本章学习使用 Spring JDBC 来简化原始的 JDBC 操作数据库，介绍 JdbcTemplate 模板及其提供的各种方法与应用，并介绍数据源的配置方法。

4.1 Spring JDBC 简介

采用原始的 JDBC 操作数据库步骤繁多，代码冗长复杂。Spring 框架为解决 Java EE 开发的复杂性，简化了众多的 Java EE API，其中 Spring JDBC 模块大大简化了 JDBC 的操作。Spring JDBC 模块提供了一个关键类 JdbcTemplate，它封装了操作数据库的多个实用方法，获取 JdbcTemplate 类的实例化对象后就可以调用这些方法操作数据库，简单、快捷。JdbcTemplate 类还有一个重要属性 DataSource，用于获取数据库连接，支持连接池和分布式事务，只有先建立好了数据库连接，JdbcTemplate 才能发挥作用。项目中使用 Spring JDBC 的基本流程如下。

（1）在 Spring 配置文件中配置好数据源 DataSource，创建 id 为 DataSource 的 Bean。一般情况下，数据源使用的类是 org.springframework.jdbc.datasource.DriverManagerDataSource。也可以采用支持连接池的 DBCP 数据源或 C3P0 数据源。

（2）在配置文件中将步骤（1）创建的 id 为 dataSource 的 Bean 注入给 id 为 jdbcTemplate 的 JdbcTemplate 类的 Bean。

（3）在配置文件中将步骤（2）创建的 id 为 jdbcTemplate 的 Bean 注入给引用了它的 DAO 层类的 jdbcTemplate 属性。

4.2 JdbcTemplate 各种方法的使用

视频讲解

JdbcTemplate 提供了 execute 方法、update 方法、query 方法等，可以满足常用的增、删、改、查操作需求，各个方法又有多种重载或变形，下面进行详细介绍。

4.2.1 execute 方法

execute 方法的语法形式是 void execute(String sql)，可以用来执行各种 SQL 语句，但由于没有返回值，用于执行 select 查询就不太合适，通常用于执行 SQL 中的 DDL，如创建数据库中的表。

示例： 用 Spring 操作数据库 goods，实现对数据的增、删、改、查操作。本节首先实现动态创建数据库表 goodsinfo 的功能。

实现步骤：

（1）在 MySQL 数据库中新建数据库 goods（无须建表，将由 Java 程序动态创建）。

（2）在 Eclipse 中新建项目 goods，导入 Spring 所需的 JAR 包，包括 Spring JDBC 的 JAR 包（spring-jdbc-4.3.4.RELEASE.jar）、事务处理的 JAR 包（spring-tx-4.3.4.RELEASE.jar）和连接 MySQL 数据库所需的 JAR 包。最终导入的 JAR 包如图 4-1 所示。

图 4-1　Spring 所需的 JAR 包

（3）在项目的 src 目录下创建 com.entity 包，包下创建 GoodsInfo 实体类，关键内容如下所示。

```
public class GoodsInfo {
    private int id;
    private String goodsName;
    private double price;
    private int quantity;
    //省略 getter、setter 方法和构造方法
}
```

（4）在项目的 src 目录下创建 com.dao 包，用作数据访问层；包下创建接口 GoodsDao，在该接口中创建方法 create，用来创建数据库表，代码如下。

```
public interface GoodsDao {
    public void create();
}
```

（5）在 com.dao 包下创建上述接口的实现类 GoodsDaoImpl。

```
public class GoodsDaoImpl implements GoodsDao{
    private JdbcTemplate jdbcTemplate;
    //省略 getter 和 setter 方法
    @Override
    public void create() {
        Stringsql="create table goodsinfo(id int primary key auto_increment,
```

```
goodsname
            varchar(50),price double,quantity int)";
            jdbcTemplate.execute(sql);
            System.out.println("数据库商品信息表 goodsinfo 创建成功!");
        }
    }
```

上述代码中的属性 jdbcTemplate 并没有直接实例化，而是通过 IoC 方式注入，具体见步骤（6）。create 方法中的 sql 是创建数据表的 DDL 语句，然后调用 jdbcTemplate 的 execute 方法执行。

（6）在项目的 src 目录下创建 Spring 配置文件 applicationContext.xml，并按 4.1 节的描述先后创建三个 Bean:dataSource 数据源、jdbcTemplate、数据访问层 goodsDao，代码如下所示。

```
    <?xml version="1.0" encoding="UTF-8"?>
    <beans xmlns="http://www.springframework.org/schema/beans"
        xmlns:xsi="http://www.w3.org/2001/XMLSchema-instance"
        xmlns:aop="http://www.springframework.org/schema/aop"
        xmlns:context="http://www.springframework.org/schema/context"
        xsi:schemaLocation="
            http://www.springframework.org/schema/beans
            http://www.springframework.org/schema/beans/spring-beans.xsd
            http://www.springframework.org/schema/context
            http://www.springframework.org/schema/context/spring-context.xsd
            http://www.springframework.org/schema/aop
            http://www.springframework.org/schema/aop/spring-aop.xsd">
        <!-- 配置数据源，创建第一个 Bean:dataSource -->
        <bean id="dataSource" class="org.springframework.jdbc.datasource.
DriverManagerDataSource">
            <property name="driverClassName">
                <value>com.mysql.jdbc.Driver</value>
            </property>
            <property name="url">
                <value>jdbc:mysql://localhost:3306/goods </value>
            </property>
            <property name="username">
                <value>root</value>
            </property>
            <property name="password">
                <value>root</value>
            </property>
        </bean>
        <!-- 配置 JdbcTemplate 模板,注入 dataSource,创建第二个 Bean: jdbcTemplate-->
        <bean id="jdbcTemplate" class="org.springframework.jdbc.core.
JdbcTemplate">
            <property name="dataSource" ref="dataSource" />
```

```
    </bean>
    <!-- 配置 DAO,注入 jdbcTemplate,创建第三个 Bean: goodsDao -->
    <bean id="goodsDao" class="com.dao.GoodsDaoImpl">
        <property name="jdbcTemplate" ref="jdbcTemplate"/>
    </bean>
</beans>
```

（7）在项目的 src 目录下创建包 com.test，包下创建 Test 类用于测试，该类下创建方法 create，代码如下所示。

```
public static void main(String[] args) {
        createDB();
    }
    private static void createDB() {
        ApplicationContext context=new
ClassPathXmlApplicationContext("applicationContext.xml");
        GoodsDao goodsDao=(GoodsDao) context.getBean("goodsDao");
        System.out.println("----------创建数据库表 goodsinfo---------");
        goodsDao.create();
    }
```

（8）运行测试类，控制台输出如下。

```
----------创建数据库表 goodsinfo---------
数据库商品信息表 goodsinfo 创建成功!
```

刷新数据库，发现多了一张表 goodsinfo，说明测试成功。

4.2.2 update 方法

JdbcTemplate 类的 update 方法用于数据库的添加、删除和修改操作。该方法有多个重载，适合各种灵活多样的输入参数，常用的重载方法主要有两个，具体如表 4-1 所示。

表 4-1 JdbcTemplate 类的 update 方法

重 载 方 法	描 述
int update(String sql)	根据参数 sql 中完整的增加、删除和修改语句执行数据操作，返回受影响行数
Int update(String sql, Object…args)	参数 sql 通常是带“?”占位符参数的，args 用于为这些占位符参数赋值。同样地，返回受影响行数

示例：用 Spring 操作数据库 goods，实现对数据的增、删、改、查操作。本节实现增加、删除和修改功能。

实现步骤：

（1）在包 com.dao 下的 GoodsDao 接口中添加以下方法。

```
public int add(GoodsInfo goodsInfo);
public int delete(int id);
public int update(GoodsInfo goodsInfo);
```

（2）在包 com.dao 下的 GoodsDaoImpl 类中添加以下代码，分别实现添加商品、修改商品和删除商品的功能。

```
    private JdbcTemplate jdbcTemplate;
    public JdbcTemplate getJdbcTemplate() {
        return jdbcTemplate;
    }
    public void setJdbcTemplate(JdbcTemplate jdbcTemplate) {
        this.jdbcTemplate = jdbcTemplate;
    }

    //添加商品
    @Override
    public int add(GoodsInfo goodsInfo) {
         Stringsql="insert into goodsinfo(goodsname,price,quantity) values
(?,?,?)";
         Object[] params={goodsInfo.getGoodsName(),goodsInfo.getPrice(),
goodsInfo.getQuantity()};
         return jdbcTemplate.update(sql,params);
    }

    //删除商品
    @Override
    public int delete(int id) {
         String sql="delete from goodsinfo where id=?";
         return jdbcTemplate.update(sql,id);
    }
    //修改商品
    @Override
    public int update(GoodsInfo goodsInfo) {
         Stringsql="update goodsinfo set goodsname=?,price=?,quantity=? where
id=?";
         Object[] params={goodsInfo.getGoodsName(),goodsInfo.getPrice(),
goodsInfo.getQuantity(),goodsInfo.getId()};
         return jdbcTemplate.update(sql,params);
    }
```

（3）在测试类 Test.java 中添加以下方法。

```
    public static void main(String[] args) {
         //CreateDB();
         addGoods();
         updateGoods();
         deleteGoods();
    }
    //添加商品
    private static void addGoods() {
```

```
        ApplicationContext context=new ClassPathXmlApplicationContext
    ("applicationContext.xml");
        GoodsDao goodsDao=(GoodsDao) context.getBean("goodsDao");
        System.out.println("----------添加一个商品信息---------");
        Scanner input=new Scanner(System.in);
        System.out.print("请输入商品名称:");
        String goodsname=input.next();
        System.out.print("请输入商品价格:");
        double price=input.nextDouble();
        System.out.print("请输入商品数量:");
        int quantity =input.nextInt();
        GoodsInfo goodsinfo=new GoodsInfo(goodsname,price,quantity);
        int count=goodsDao.add(goodsinfo);
        if(count>0) {
            System.out.println("添加商品信息成功!");
        }else {
             System.out.println("添加商品信息失败!");
        }
    }
    //修改商品
    private static void updateGoods() {
        ApplicationContext context=new ClassPathXmlApplicationContext
    ("applicationContext.xml");
        GoodsDao goodsDao=(GoodsDao) context.getBean("goodsDao");
        System.out.println("----------修改商品信息---------");
        Scanner input=new Scanner(System.in);
        System.out.print("请输入要修改的商品编号:");
        int id=input.nextInt();
        System.out.print("请输入新商品名称:");
        String goodsname=input.next();
        System.out.print("请输入新商品价格:");
        double price=input.nextDouble();
        System.out.print("请输入新商品数量:");
        int quantity =input.nextInt();
        GoodsInfo goodsinfo=new GoodsInfo(id,goodsname,price,quantity);
        int count=goodsDao.update(goodsinfo);
        if(count>0) {
            System.out.println("修改商品信息成功!");
        }else {
             System.out.println("修改商品信息失败!");
        }
    }
    //删除商品
    private static void deleteGoods() {
        ApplicationContext context=new
    ClassPathXmlApplicationContext("applicationContext.xml");
        GoodsDao goodsDao=(GoodsDao) context.getBean("goodsDao");
```

```
        System.out.println("----------删除商品信息---------");
        Scanner input=new Scanner(System.in);
        System.out.print("请输入要删除的商品编号:");
        int id=input.nextInt();
        int count=goodsDao.delete(id);
        if(count>0) {
            System.out.println("删除商品信息成功!");
        }else {
            System.out.println("删除商品信息失败!");
        }
    }
```

（4）运行测试结果如下。

```
----------添加一个商品信息---------
请输入商品名称:iPhone8
请输入商品价格:10000
请输入商品数量:10
添加商品信息成功!
----------修改商品信息---------
请输入要修改的商品编号:
1
请输入新商品名称:iPhone9
请输入新商品价格:20000
请输入新商品数量:20
修改商品信息成功!
----------删除商品信息---------
请输入要删除的商品编号:1
删除商品信息成功!
```

至此，商品的增加、删除和修改成功!

4.2.3 query 方法

query 方法有多个重载，此外还有变形，详见表 4-2 的说明。

表 4-2 JdbcTemplate 的 query 方法及其变形

重载方法	描　　述

List<T> query(String sql, RowMapper<T> rowMapper)	执行不带参数的 select 语句，返回多条值的情况，封装成 T 类型的泛型集合，事先要定义好 RowMapper<T>对象 rowMapper。示例：查询所有商品信息。
	`RowMapper<GoodsInfo> rowMapper=new BeanPropertyRowMapper<GoodsInfo>(GoodsInfo.class);`
	`List<GoodsInfo> list=jdbcTemplate.query(sql, rowMapper);`

续表

重载方法	描　述
List<T> query(String sql,Object[] args,RowMapper<T> rowMapper)	执行带参数的 select 语句，返回多条值的情况，封装成 T 类型的泛型集合,事先要定义好 RowMapper<T>对象 rowMapper。示例：查询价格大于 100 元的商品。 RowMapper<GoodsInfo> rowMapper=**new** BeanPropertyRowMapper<GoodsInfo>(GoodsInfo.**class**); List<GoodsInfo> list=jdbcTemplate.query(sql, params, rowMapper);
SqlRowSet queryForRowSet(String sql)	可用于查询部分列或类似 count(*)的聚合查询语句，返回 SqlRowSet 行集合。需要调用 next 方法移到行集合的第一行，再用 getInt 获取列号。示例：查询商品的总数。 String sql="select count(*) from goodsinfo"; SqlRowSet rs=jdbcTemplate.queryForRowSet(sql); rs.next(); **int** count=rs.getInt(1);
T queryForObject(String sql, RowMapper<T> rowMapper)	执行不带参数的 select 语句，返回单条值的情况，封装成 T 类型,事先要定义好 RowMapper<T>对象 rowMapper。示例：查询单条商品。 RowMapper<GoodsInfo> rowMapper=**new** BeanPropertyRowMapper<GoodsInfo>(Student.**class**); GoodsInfo student=jdbcTemplate.queryForObject(sql, rowMapper);
T queryForObject(String sql, Object[] args,RowMapper<T> rowMapper)	执行带参数的 select 语句，返回单条值的情况，封装成 T 类型，事先要定义好 RowMapper<T>对象 rowMapper。示例：查询 id 为某个给定值的单个商品。 RowMapper<GoodsInfo> rowMapper=**new** BeanPropertyRowMapper<GoodsInfo>(GoodsInfo.**class**); GoodsInfo goodsinfo=jdbcTemplate.queryForObject(sql, params,rowMapper);

示例：用 Spring 操作数据库 goods，实现对数据的增、删、改、查操作。本节实现查询所有商品，查询价格大于 15000 元的商品，查询商品种类数量，查询单条商品，查询 id 为某个给定值的单个商品的功能。

实现步骤:

（1）在包 com.dao 下的 GoodsDao 接口中添加以下方法。

```
public List<GoodsInfo> findAllGoods();  //查询所有商品
public List<GoodsInfo> findGoods(double price);//查询价格大于某个值的商品
public int getGoodsCount();             //查询商品种类数量
public GoodsInfo findSingleGoods();     //查询单条商品
public GoodsInfo findGoodsById(int id); //查询某个 id 号的商品
```

（2）在包 com.dao 下的 GoodsDaoImpl 类中添加以下方法，分别实现查询价格大于 15000 元的商品，查询商品种类数量，查询单条商品，查询 id 为某个给定值的单个商品的功能。

```
//查询所有商品
    @Override
    public List<GoodsInfo> findAllGoods() {
        String sql="select * from goodsinfo";
        RowMapper<GoodsInfo> rowMapper=new
        BeanPropertyRowMapper<GoodsInfo>(GoodsInfo.class);
        List<GoodsInfo> list=jdbcTemplate.query(sql, rowMapper);
        return list;
    }

    //查询价格大于某个值的商品
    @Override
    public List<GoodsInfo> findGoods(double price) {
        String sql="select * from goodsinfo where price>?";
        Object[] params={price};
        RowMapper<GoodsInfo> rowMapper=new
        BeanPropertyRowMapper<GoodsInfo>(GoodsInfo.class);
        List<GoodsInfo> list=jdbcTemplate.query(sql, params,rowMapper);
        return list;
    }

    //查询商品种类数量
    @Override
    public int getGoodsCount() {
        String sql="select count(*) from goodsinfo";
        SqlRowSet rs=jdbcTemplate.queryForRowSet(sql);
        int count=rs.getInt(0);
        return count;
    }

    //查询单条商品
    @Override
    public GoodsInfo findSingleGoods() {
        String sql="select * from goodsinfo where id=1";
        RowMapper<GoodsInfo> rowMapper=new
        BeanPropertyRowMapper<GoodsInfo>(GoodsInfo.class);
        GoodsInfo goodsInfo=jdbcTemplate.queryForObject(sql,rowMapper);
        return goodsInfo;
    }

    //查询某个 id 的商品
    @Override
    public GoodsInfo findGoodsById(int id) {
        String sql="select * from goodsinfo where id=?";
        Object[] params={id};
        RowMapper<GoodsInfo> rowMapper=new
        BeanPropertyRowMapper<GoodsInfo>(GoodsInfo.class);
```

```
        GoodsInfo goodsInfo=jdbcTemplate.queryForObject(sql, params,rowMapper);
        return goodsInfo;
    }
```

（3）在 com.test 包下的 Test 类中创建如下方法，分别测试上述功能。

```
public static void main(String[] args) {
        findAllGoods();
        findGoods();
        getGoodsCount();
        findSingleGoods();
        findGoodsById();
    }
//查询所有商品
    private static void findAllGoods() {
        ApplicationContext context=new
        ClassPathXmlApplicationContext("applicationContext.xml");
        GoodsDao goodsDao=(GoodsDao) context.getBean("goodsDao");
        System.out.println("----------查询所有商品---------");
        List<GoodsInfo> list=goodsDao.findAllGoods();
        System.out.println("商品 id\t 商品名称\t 商品价格\t 商品数量");
        for(GoodsInfo goodsinfo:list) {
        System.out.println(goodsinfo.getId()+"\t"+goodsinfo.getGoodsName()
            +"\t"+goodsinfo.getPrice()+"\t"+goodsinfo.getQuantity());
        }
    }
    //查询价格大于 15000 元的商品
    private static void findGoods() {
        ApplicationContext context=new
        ClassPathXmlApplicationContext("applicationContext.xml");
        GoodsDao goodsDao=(GoodsDao) context.getBean("goodsDao");
        System.out.println("----------查询价格大于 15000 元的商品---------");
        List<GoodsInfo> list=goodsDao.findGoods(15000.00);
        System.out.println("商品 id\t 商品名称\t 商品价格\t 商品数量");
        for(GoodsInfo goodsinfo:list) {
        System.out.println(goodsinfo.getId()+"\t"+goodsinfo.getGoodsName()
           +"\t"+goodsinfo.getPrice()+"\t"+goodsinfo.getQuantity());
        }
    }
    //查询商品种类数量
    private static void getGoodsCount() {
        ApplicationContext context=new
        ClassPathXmlApplicationContext("applicationContext.xml");
        GoodsDao goodsDao=(GoodsDao) context.getBean("goodsDao");
        System.out.println("----------查询商品种类数量---------");
        int count=goodsDao.getGoodsCount();
        System.out.println("商品种类数量:"+count);
```

```
    }
    //查询单条商品
    private static void findSingleGoods() {
        ApplicationContext context=new
        ClassPathXmlApplicationContext("applicationContext.xml");
        GoodsDao goodsDao=(GoodsDao) context.getBean("goodsDao");
        System.out.println("----------查询单条商品---------");
        GoodsInfo goodsinfo=goodsDao.findSingleGoods();
        System.out.println("商品id\t商品名称\t商品价格\t商品数量");
        System.out.println(goodsinfo.getId()+"\t"+goodsinfo.getGoodsName()
        +"\t"+goodsinfo.getPrice()+"\t"+goodsinfo.getQuantity());
    }
    //查询某个id的商品
    private static void findGoodsById() {
        ApplicationContext context=new
        ClassPathXmlApplicationContext("applicationContext.xml");
        GoodsDao goodsDao=(GoodsDao) context.getBean("goodsDao");
        System.out.println("----------查询单条商品---------");
        GoodsInfo goodsinfo=goodsDao.findGoodsById(2);
        System.out.println("商品id\t商品名称\t商品价格\t商品数量");
        System.out.println(goodsinfo.getId()+"\t"+goodsinfo.getGoodsName()
            +"\t"+goodsinfo.getPrice()+"\t"+goodsinfo.getQuantity());
    }
```

（4）运行测试类，结果如下所示。

```
----------查询所有商品---------
商品id    商品名称    商品价格    商品数量
1    iPhone9    10000.0    30
2    iPhone8    20000.0    20
----------查询价格大于15000元的商品---------
商品id    商品名称    商品价格    商品数量
2    iPhone8    20000.0    20
----------查询商品种类数量---------
商品种类数量:2
----------查询单条商品---------
商品id    商品名称    商品价格    商品数量
1    iPhone9    10000.0    30
----------查询单条商品---------
商品id    商品名称    商品价格    商品数量
2    iPhone8    20000.0    20
```

4.3 数据源的配置

Spring有3种数据源可以选择，4.2节的案例在操作数据库时用到了其中的默认数据源。Spring的3种数据源分别是：

（1）默认数据源 DriverManagerDataSource；

（2）DBCP 数据源；

（3）C3P0 连接池数据源。

视频讲解

4.3.1 DBCP 数据源 BasicDataSource 的使用

使用 DBCP 数据源前需要导入以下两个 JAR 包。

（1）commons-dbcp-osgi-1.2.2.jar；

（2）commons-pool-1.5.3.jar。

示例：用 Spring 操作数据库 goods,实现对数据的增、删、改、查操作,使用 DBCP 数据源。

实现步骤：

（1）复制项目 goods 为 goods2，再添加上述两个 JAR 包。

（2）修改配置文件如下。

```
<?xml version="1.0" encoding="UTF-8"?>
<beans xmlns="http://www.springframework.org/schema/beans"
   xmlns:xsi="http://www.w3.org/2001/XMLSchema-instance"
   xmlns:aop="http://www.springframework.org/schema/aop"
   xmlns:context="http://www.springframework.org/schema/context"
   xsi:schemaLocation="
      http://www.springframework.org/schema/beans
      http://www.springframework.org/schema/beans/spring-beans.xsd
      http://www.springframework.org/schema/context
      http://www.springframework.org/schema/context/spring-context.xsd
      http://www.springframework.org/schema/aop
      http://www.springframework.org/schema/aop/spring-aop.xsd">
   <!-- 配置数据源 -->
   <bean id="dataSource" class="org.apache.commons.dbcp.BasicDataSource">
      <property name="driverClassName">
         <value>com.mysql.jdbc.Driver</value>
      </property>                                  这里改用了 DBCP 数据源
      <property name="url">
         <value>jdbc:mysql://localhost:3306/goods</value>
      </property>
      <property name="username">
         <value>root</value>
      </property>
      <property name="password">
         <value>root</value>
      </property>
   </bean>
   <!-- 配置 JdbcTemplate 模板 -->
   <bean id="jdbcTemplate" class="org.springframework.jdbc.core.
JdbcTemplate">
```

```
        <property name="dataSource" ref="dataSource" />
    </bean>
    <!-- 配置 DAO 层,注入 jdTemplate 属性值 -->
    <bean id="goodsDao" class="com.dao.GoodsDaoImpl">
        <property name="jdbcTemplate" ref="jdbcTemplate"/>
    </bean>
</beans>
```

（3）其他地方不变，运行测试类，数据操作的结果一样，说明数据源配置成功。

4.3.2 C3P0 数据源 ComboPooledDataSource 的使用

视频讲解

使用 C3P0 数据源前需要导入 JAR 包 c3p0-0.9.1.2.jar，并且配置关键字也有不同之处，详见下面的项目案例。

示例：用 Spring 操作数据库 goods,实现对数据的增、删、改、查操作,使用 DBCP 数据源。

实现步骤：

（1）复制项目 goods2 为 goods3，添加上述 JAR 包。

（2）修改配置文件如下。

```
<?xml version="1.0" encoding="UTF-8"?>
<beans xmlns="http://www.springframework.org/schema/beans"
    xmlns:xsi="http://www.w3.org/2001/XMLSchema-instance"
    xmlns:aop="http://www.springframework.org/schema/aop"
    xmlns:context="http://www.springframework.org/schema/context"
    xsi:schemaLocation="
        http://www.springframework.org/schema/beans
        http://www.springframework.org/schema/beans/spring-beans.xsd
        http://www.springframework.org/schema/context
        http://www.springframework.org/schema/context/spring-context.xsd
        http://www.springframework.org/schema/aop       这里改用了 C3P0 数据源
        http://www.springframework.org/schema/aop/spring-aop.xsd">
    <!-- 配置数据源 -->
    <bean id="dataSource" class="com.mchange.v2.c3p0.ComboPooledDataSource">
        <property name="driverClass">
            <value>com.mysql.jdbc.Driver</value>  注意加粗部分属性名称不同
        </property>
        <property name="jdbcUrl">
            <value>jdbc:mysql://localhost:3306/goods</value>
        </property>
        <property name="user">
            <value>root</value>
        </property>
        <property name="password">
            <value>root</value>
        </property>
```

```
    </bean>
    <!-- 配置 JdbcTemplate 模板 -->
    <bean id="jdbcTemplate" class="org.springframework.jdbc.core.
JdbcTemplate">
        <property name="dataSource" ref="dataSource" />
    </bean>
    <!-- 配置 DAO 层,注入 jdTemplate 属性值 -->
    <bean id="goodsDao" class="com.dao.GoodsDaoImpl">
        <property name="jdbcTemplate" ref="jdbcTemplate"/>
    </bean>
</beans>
```

（3）运行测试类，数据操作的结果一样，说明数据源配置成功。

4.3.3 使用属性文件读取数据库连接信息

为了方便移植到不同的数据库，如果 Spring 配置文件中有关数据库的连接信息需要能够独立出来，单独成为一个文件，并以适当方式供 Spring 配置文件调用。的确可以这样实现，数据库连接信息可以单独放入一个 properties 文件中，供 Spring 配置文件调用，若要移植到不同的数据库，只需要修改这个文件，或者 Spring 配置文件调用另外一个 properties 文件即可实现。

示例：用 Spring 操作数据库 goods，实现对数据的增、删、改、查操作，使用 C3P0 数据源，数据库连接信息存放于 src 下独立的 jdbc.properties 文件中。

实现步骤：

（1）复制项目 goods3 为 goods4，在 src 下添加属性文件 jdbc.properties，代码如下。

```
jdbc.driver=com.mysql.jdbc.Driver
jdbc.url=jdbc:mysql://localhost:3306/goods
jdbc.username=root
jdbc.password=root
```

（2）修改配置文件，先要注册属性文件，再在数据源配置中用 EL 表达式引用属性文件。

```
<?xml version="1.0" encoding="UTF-8"?>
<beans xmlns="http://www.springframework.org/schema/beans"
    xmlns:xsi="http://www.w3.org/2001/XMLSchema-instance"
    xmlns:aop="http://www.springframework.org/schema/aop"
    xmlns:context="http://www.springframework.org/schema/context"
    xsi:schemaLocation="
        http://www.springframework.org/schema/beans
        http://www.springframework.org/schema/beans/spring-beans.xsd
        http://www.springframework.org/schema/context
        http://www.springframework.org/schema/context/spring-context.xsd
        http://www.springframework.org/schema/aop
        http://www.springframework.org/schema/aop/spring-aop.xsd">
    <!-- 注册属性文件第一种方式 -->
```

```
        <bean class="org.springframework.beans.factory.config.
PropertyPlaceholderConfigurer">
            <property name="location" value="classpath:jdbc.properties"/>
        </bean>
        <!-- 配置数据源 -->
        <bean id="dataSource" class="com.mchange.v2.c3p0.ComboPooledDataSource">
            <property name="driverClass">
                <value>${jdbc.driver}</value>
            </property>
            <property name="jdbcUrl">
                <value>${jdbc.url}</value>
            </property>
            <property name="user">
                <value>${jdbc.username}</value>
            </property>
            <property name="password">
                <value>${jdbc.password}</value>
            </property>
        </bean>
        <!-- 配置 JdbcTemplate 模板 -->
        <bean id="jdbcTemplate" class="org.springframework.jdbc.core.
JdbcTemplate">
            <property name="dataSource" ref="dataSource" />
        </bean>
        <!-- 配置 DAO 层,注入 jdbcTemplate 属性值 -->
        <bean id="goodsDao" class="com.dao.GoodsDaoImpl">
            <property name="jdbcTemplate" ref="jdbcTemplate"/>
        </bean>
    </beans>
```

注册属性文件可以使用上述代码中用到的方式。

```
    <!-- 注册属性文件第一种方式 -->
        <bean class="org.springframework.beans.factory.config.
PropertyPlaceholderConfigurer">
            <property name="location" value="classpath:jdbc.properties"/>
        </bean>
```

除此之外，还有另一种方式。

```
    <!-- 注册属性文件第二种方式 -->
        <context:property-placeholder location="classpath:jdbc.properties"/>
```

这种方式要求配置文件头要有 context 约束信息。

两种方式的效果相同，选用其中一种即可。考虑到第二种方式相对简单些，本案例选用第二种。

（3）运行测试类，可以发现两种方式的读取信息效果一样。

4.4 本章小结 

本章介绍了 Spring 框架自带的 JDBC 模块的使用方法。Spring JDBC 模块是在原生 JDBC 基础上的增强模块，提供了很多便利功能。利用 JdbcTemplate 模板类，简化了常规的 JDBC 操作，在不使用 MyBatis 框架的情况下，也不失为一个好的技术方案。

习题 4

1. 使用 Spring JDBC 的基本流程是什么？
2. Spring JDBC 的 JdbcTemplate 类主要方法有哪些？
3. JdbcTemplate 类的 query 方法有哪些重载？各有什么功能？
4. 利用 Spring JDBC 在数据库中创建数据表 employee，包含员工编号 empno、姓名 empname、性别 gender、职位 position、工资 salary、所在部门 department 等。使用 Spring JDBC 添加若干新员工，查询所有员工的信息，查询某个部门的员工信息，查询女员工的信息，查询员工总人数，查询工资大于某个值的员工，查询某个 empno 的员工，修改某个员工的信息，删除某个员工的信息，并再次进行查询以确认操作效果。

第5章 Spring MVC

本章学习内容

- Spring MVC 简介；
- Spring MVC 程序运行原理；
- Spring MVC 的体系结构；
- 基于注解方式配置控制器；
- Spring MVC 注解详解。

5.1 Spring MVC 简介

前面的章节介绍了 Spring IoC、Spring AOP 和 Spring 对 JDBC 的支持。Spring 框架可以说是企业开发中技术体系最全面的一个框架，围绕着 Spring Framework 已经形成了完整的技术体系。目前，基于 Web 的 MVC 框架非常多，发展也很快，如 JSF、Struts 1、Struts 2 和 Spring MVC 等。Struts 1.x 框架在 2001 年出现后成为主流，之后陆续出现了 Struts 2 和 JSF 等框架。从架构上说，Struts 2 是一款非常优秀的软件，几乎成为 Java EE 开发的 MVC 框架的事实标准。但是由于 Struts 2 团队基本不再对其进行更新，只发布补丁，随着时间的推移，特别是 2010 年之后，Struts 2 陆续曝出了多个漏洞。很多网络安全和开发团队都给出了类似的建议：鉴于 Struts 2 至今为止已经多次曝出严重的高危漏洞，如果不是必要，建议开发者以后考虑采用其他类似的 Java 开发框架。所以广大的开发者将目光移向了 Spring MVC。Spring MVC 是后起之秀，从应用上来说要复杂一些，但是它基于 Spring 进行开发，继承了 Spring 的优秀血脉，所以一跃成为采用率最高的 Java EE Web MVC 框架。

视频讲解

5.2 第一个 Spring MVC 案例

首先通过一个简单的案例来了解 Spring MVC。

该案例的实现步骤如下。

（1）获取 Spring 框架的 JAR 库文件。

截至 2020 年 5 月，Spring Framework 已发布的稳定版本（GA 版）为 Spring 5.2.6 版，如图 5-1 所示。Spring 官方网站改版后，建议通过 Maven 和 Gradle 下载，若不使用 Maven 和 Gradle 开发项目，可以通过 Spring Framework 官网直接下载，下载路径为 http://repo.springsource.org/libs-release-local/org/springframework/spring。

对于已下载的 ZIP 文件，可以直接解压缩在本地文件夹。由于 Spring 框架的 JAR 包的深度特别深，造成解压后的文件夹名字超过了 WinRAR 或 WinZIP 软件的最长允许范围，在 Windows 操作系统下可能会导致解压失败，如果解压过程中出现问题，可以使用 7ZIP 软件来解压。

解压完成后，在文件夹中找到 libs 文件夹，有 63 个后缀名为 jar 的文件。仔细观察可以发现，每个主题有 3 个文件，例如，核心包文件为 spring-core-5.2.26.RELEASE.jar、 spring-core--5.2.26.RELEASE-javadoc.jar 和 spring-core--5.2.26.RELEASE-sources.jar。其中文件名结尾为 javadoc 的文件是 Java 根据注释自动生成的文档，而文件名结尾为 sources 的文件则是源代码文件。可以将所有的 JAR 文件导入到 Web 项目中，最简单的方法是直接复制到项目的 WebRoot 的 lib 文件夹下。

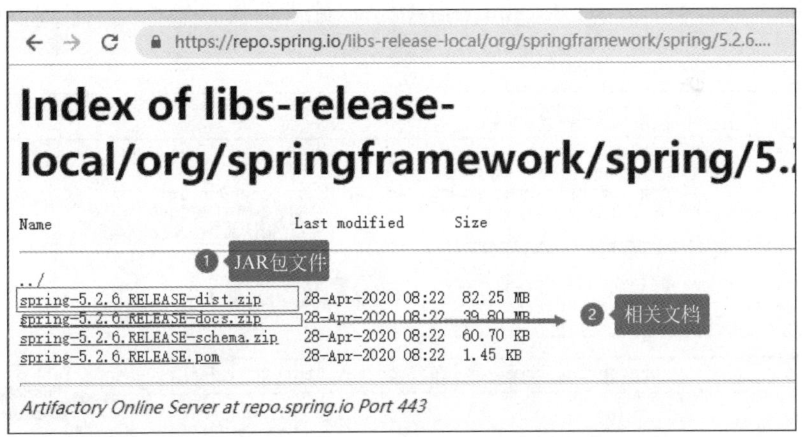

图 5-1　Spring 框架的下载

如果使用 MyEclipse2014 开发工具，可以使用 MyEclipse 内置的 Spring 包，只是版本稍低一些，不过并不影响学习和使用。Spring 5.x 的主要升级在于集成了 Spring Boot 模块，在启动方式上有很大的差别。对于初学者来说 Spring 3.x 更容易上手，所以本书后面依然采用 MyEclipse2014+Spring3.x 的方式来入门。虽然工具和版本更新换代会很快，但是 Spring 框架的思想和原理是稳定的，熟练掌握之后可以很快地完成版本切换。

（2）在 Web 项目中加入 Spring 框架。

通过 MyEclipse2014 新建一个 Web 项目，项目名为 ssmBook_ch6。在新建的 Web 项目中加入 Spring 框架，同时选中 Spring Web 模块以便支持 Spring MVC，如图 5-2 所示。

Install Spring Facet
Add Spring Capabilities
Configure Spring project
Configure Spring libraries
Type: MyEclipse Library
Libraries will be provided through MyEclipse Library System
Spring Framework version: 3.1.1
Spring Security version: <none>
Spring Webflow version: <none>
☑ Add Spring-Web support (TLDs copied to project, ContextLoaderListener added to web.xml)
☑ Enable AOP Builder
Larger projects that do not use Spring AOP may see a performance gain by leaving this disabled.
☑ Specify new or existing Spring bean configuration file?
Bean configuration type: ◉ New ○ Existing
Folder: src Browse
File: applicationContext.xml

图 5-2　在 MyEclipse 中添加 Spring MVC

（3）配置 Dispatchservlet 和对应的 Servlet。

接下来在项目的 web.xml 配置文件中加入 DispatchServlet 的配置。

Servlet 的名字在这里命名为 springapp，也可以自由定义，但是后续的定义都要修改成对应自定义的名字，如图 5-3 所示。由于 Spring MVC 采用了约定优先于配置的方式，它会将此处的 Servlet 名字加上-servlet.xml 的后缀，以此来查找一个名为 springapp-servlet.xml 的配置文件。

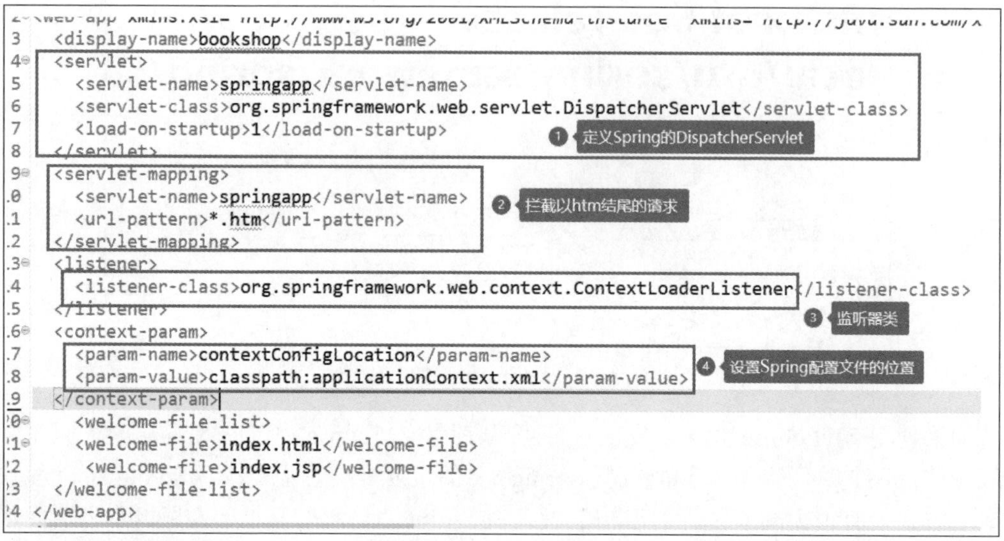

```xml
<web-app xmlns:xsi="http://www.w3.org/2001/XMLSchema-instance" xmlns="http://java.sun.com/x
  <display-name>bookshop</display-name>
  <servlet>
    <servlet-name>springapp</servlet-name>
    <servlet-class>org.springframework.web.servlet.DispatcherServlet</servlet-class>
    <load-on-startup>1</load-on-startup>
  </servlet>
  <servlet-mapping>
    <servlet-name>springapp</servlet-name>
    <url-pattern>*.htm</url-pattern>
  </servlet-mapping>
  <listener>
    <listener-class>org.springframework.web.context.ContextLoaderListener</listener-class>
  </listener>
  <context-param>
    <param-name>contextConfigLocation</param-name>
    <param-value>classpath:applicationContext.xml</param-value>
  </context-param>
    <welcome-file-list>
    <welcome-file>index.html</welcome-file>
     <welcome-file>index.jsp</welcome-file>
  </welcome-file-list>
</web-app>
```

❶ 定义Spring的DispatcherServlet
❷ 拦截以htm结尾的请求
❸ 监听器类
❹ 设置Spring配置文件的位置

图 5-3　web.xml 配置文件

在与 web.xml 文件相同的位置上，新建一个名为 springapp-servlet.xml 的 xml 文件。该 springapp-servlet.xml 配置文件中定义了用户请求的路径和对应的控制器的映射关系，如图 5-4 所示。

```xml
<!-- 定义用户请求路径和对应的响应处理类之间的关系 -->
 <bean name="/hello.htm"
 class="com.ssmbook2020.web.ch5.HelloController" />
```

图 5-4　springapp-servlet.xml 配置文件

在这个配置文件中，核心的语句是

<bean name="*/hello.htm*" class="*org.ssmbook2020.web.ch5.HelloController*"/>。

当用户请求的路径是 hello.htm 时，它告诉 spring 用对应包的 HelloController 类来处理用户的请求。该语句用来取代以前纯 servlet 开发时 servlet 和 url 之间的关系映射。

（4）创建控制器类。

接下来创建 HelloController 的类，它的作用类似于以前的 Servlet。

代码如下所示：

```java
//代码清单 5-1 HelloController.java
import org.springframework.web.servlet.ModelAndView;
import org.springframework.web.servlet.mvc.Controller;
```

```
public class HelloController implements Controller {
    //返回 ModelAndView 对象
    public ModelAndView handleRequest(HttpServletRequest request,
            HttpServletResponse response)
            throws ServletException, IOException{
        //向 Request 域中放入一条信息，给前端 jsp 用
        request.setAttribute("message", "hello,Spring MVC");
        //返回 jsp 的路径
        return new ModelAndView("hello.jsp");
    }
}
```

这个控制器类比原始的 Servlet 更简洁，它继承了 Controller 接口，并实现了 handleRequest 方法。handleRequest 方法的两个参数就是原始的 HttpServletRequest 和 HttpServletResponse。为了演示效果，使用 Request 放入一个字符串信息，然后在 hello.jsp 页面中显示出这个字符串。

（5）建立 jsp 文件。

在 WebRoot 文件夹下新建 hello.jsp 文件，其内容如下。

```
<%@ page language="java" import="java.util.*" pageEncoding="UTF-8"%>
<html>
  <head>
    <title>hello jsp 页面</title>
  </head>
  <body>
     显示服务器信息如下:${requestScope.message }
  </body>
</html>
```

该步骤的目的是显示 HelloController 类中放入的信息，可以直接用 EL 表达式显示出来。

（6）部署运行。

将项目部署到 Tomcat 服务器上，启动 Tomcat 服务器，在浏览器中输入地址 http://localhost:8080/ssmBook_ch6/hello.htm，结果如图 5-5 所示。

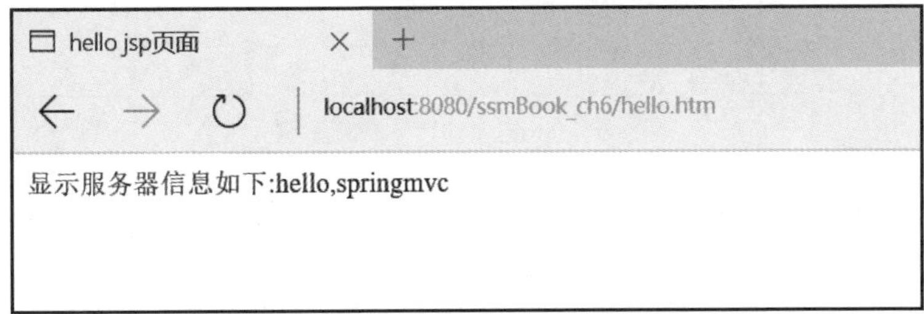

图 5-5　第一个 Spring MVC 的运行结果

5.3 Spring MVC 的工作原理与体系结构

5.3.1 Spring MVC 程序运行原理

通过第一个 Spring MVC 程序来介绍 Spring MVC 的运行流程。

（1）用户通过浏览器发出请求。

（2）在 web.xml 中 DispatchServlet 拦截*.htm 的请求。

（3）在与 web.xml 相同的路径下查找该 Servlet 对应的 Spring 配置文件，此案例中为 springapp-servlet.xml。

（4）根据 springapp-servlet.xml 配置文件中的 BeanName，找到对应的处理请求的类。在此案例中，用 HelloController 类来响应 hello.htm 请求。

（5）在 HelloController 类中的方法 `handleRequest`，它的作用类似于纯 Servlet 中的 doGet 或者 doPost。注意它的返回值是一个 Spring MVC 中的对象 ModelAndView，这个对象可以用来封装模型和视图。Hello.jsp 是默认的 jsp 页面的名字。这里直接返回页面的名字是不可取的，本章后面将会介绍更合理的方式。

图 5-6 为 Spring MVC 的工作流程图。

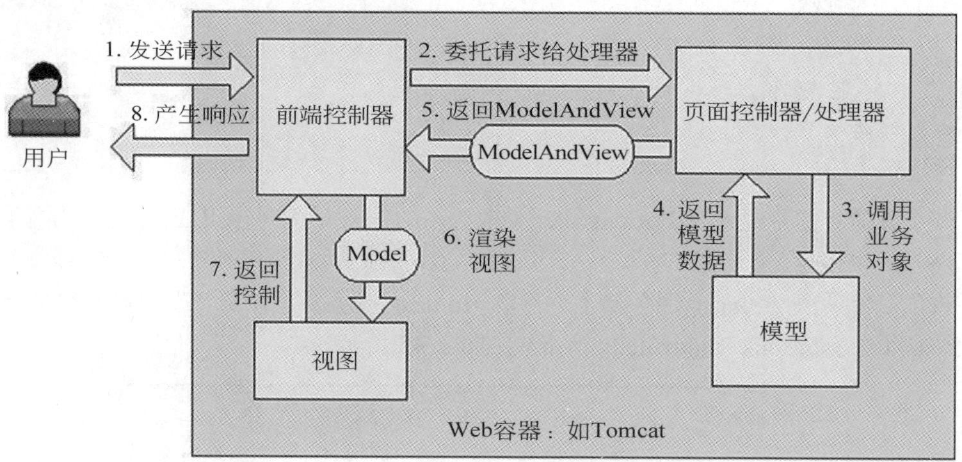

图 5-6　Spring MVC 工作流程图

5.3.2 视图解析器

在代码清单 5-1 的 HelloController 类中，handleRequest 方法返回的是一个 jsp 页面的名字。在实际案例中会改为如下的代码。

```
//代码清单 5-2  HelloController.java
package com.ssmbook2020.web.ch5;
import org.springframework.web.servlet.ModelAndView;
```

```
import org.springframework.web.servlet.mvc.Controller;

public class HelloController implements Controller {

    //返回 ModelAndView 对象
    public ModelAndView handleRequest(HttpServletRequest request,
            HttpServletResponse response)
            throws ServletException, IOException {
        //向 request 域中放入一条信息，给前端 jsp 用
        request.setAttribute("message", "hello,Spring MVC");
        //返回 jsp 的路径
        return new ModelAndView("hello");//和代码清单 5-1 的唯一区别!
    }
}
```

在代码 **return new** ModelAndView("hello")中返回的是“hello”字符串而不是“hello.jsp”的文件，为了让程序能正常工作，必须要在 springapp-servlet.xml 中加入 hello 的对应视图的解析方式，即 hello 的对应文件。Spring MVC 通过配置文件给“hello”加上前缀和后缀来指定唯一的物理文件。以下是修改后的 springapp-servlet.xml 文件内容。

```
<?xml version="1.0" encoding="UTF-8"?>
<beans xmlns="http://www.springframework.org/schema/beans"
xmlns:xsi="http://www.w3.org/2001/XMLSchema-instance"
xmlns:p="http://www.springframework.org/schema/p"
xsi:schemaLocation="http://www.springframework.org/schema/beans
http://www.springframework.org/schema/beans/spring-beans-3.1.xsd">
    <!-- 定义用户请求路径和对应的响应处理类之间的关系 -->
<bean name="/hello.htm" class="com.ssmbook2020.web.ch5.HelloController" />
<!-- 配置一个视图解析器 -->
<bean
class="org.springframework.web.servlet.view.InternalResourceViewResolver">
        <property name="prefix" value="/WEB-INF/jsp/" />
        <property name="suffix" value=".jsp" />
    </bean>
</beans>
```

其中，prefix 属性的值表示地址前缀，suffix 属性的值表示地址后缀，如图 5-7 所示。

```
<!-- 定义用户请求路径和对应的响应处理类之间的关系 -->
 <bean name="/hello.htm"
 class="com.ssmbook2020.web.ch5.HelloController" />

 <!-- 配置一个视图解析器 -->
 <bean
 class="org.springframework.web.servlet.view.InternalResourceViewResolver">
     <property name="prefix" value="/WEB-INF/jsp/" />   ❶ 路径前缀
     <property name="suffix" value=".jsp" />   ❷ 路径后缀
 </bean>
```

图 5-7　增加视图解析器

这样 hello 就会对应到 /WebRoot/WEB-INF/jsp/hello.jsp 文件，当然文件位置也要有对应变化，要把 jsp 文件放到 WEB-INF 文件夹下。这样有更好的安全性，因为用户无法直接访问 Tomcat 服务器下项目中 WEB-INF 下的文件，可以起到一定的保护作用。

5.3.3 Spring MVC 的体系结构

下面详细介绍 Spring MVC 的体系结构，图 5-8 为 Spring MVC 体系结构图。

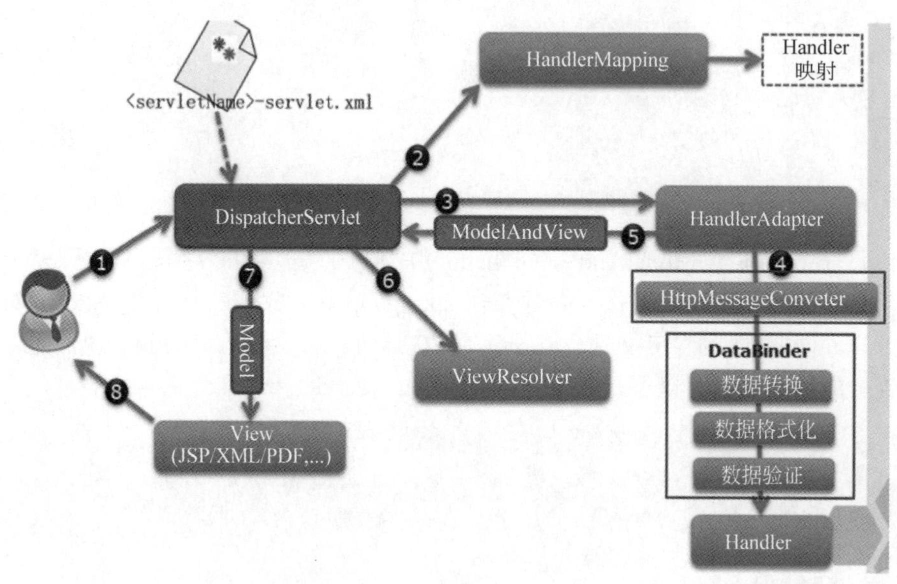

图 5-8 Spring MVC 体系结构图

分析 Spring MVC 体系结构图，可以得到整个流程如下。

（1）用户向服务器发送请求，Spring 前端控制 Servlet，请求被 DispatcherServlet 捕获。

（2）DispatcherServlet 对请求 URL 进行解析，得到请求资源标识符（URI）；然后根据该 URI 调用 HandlerMapping，获得该 Handler 配置的所有相关的对象（包括 Handler 对象和 Handler 对象对应的拦截器）；最后以 HandlerExecutionChain 对象的形式返回。

（3）DispatcherServlet 根据获得的 Handler，选择一个合适的 HandlerAdapter。如果成功获得 HandlerAdapter，此时将开始执行拦截器的 preHandler(...)方法。

（4）提取 Request 中的模型数据，填充 Handler 输入参数，开始执行 Handler(Controller)。在填充 Handler 输入参数的过程中，Spring 将根据配置做一些额外的工作。

① HttpMessageConveter：将请求消息（如 JSON、XML 等数据）转换成一个对象，将对象转换为指定的响应信息。

② 数据转换：对请求消息进行数据转换，如 String 转换成 Integer、Double 等。

③ 数据根式化：对请求消息进行数据格式化，如将字符串转换成格式化数字或格式化日期等。

④ 数据验证：验证数据的有效性（长度、格式等），将验证结果存储到 BindingResult 或 Error 中。

（5）Handler 执行完成后，向 DispatcherServlet 返回一个 ModelAndView 对象。

（6）根据返回的 ModelAndView，选择一个适合的 ViewResolver（必须是已经注册到 Spring 容器中的 ViewResolver)返回给 DispatcherServlet 。

（7）ViewResolver 结合 Model 和 View 来渲染视图。

（8）将渲染结果返回给客户端。

5.4 基于注解的控制器配置

在 5.2 节已经学习了 Spring MVC 的基本概念，对 Spring MVC 有了基本的了解。配置控制器的传统方式是采用 XML 配置形式。随着现在基于注解的方式在 Java EE 项目中逐渐流行，Spring MVC 2.5 版本后也支持基于注解的方式来配置控制器。

与 5.2 节基于 xml 配置的初始步骤相同，加入 Spring MVC 的包，并在项目的 web.xml 配置文件中加入 DispatchServlet 的配置。在与 web.xml 文件相同的位置，新建一个名为 springapp-servlet.xml 的 xml 文件。该 springapp-servlet.xml 配置文件中定义了用户请求的路径和对应的控制器的映射关系。

基于注解的控制器配置的实现步骤如下。

（1）修改控制器类

修改之前的 HelloController 类。为了和 5.2 节的 HelloController 区分，将这个类定义为 HelloController2。代码如下所示。

```
//代码清单 5-3 HelloController2.java
import org.springframework.stereotype.Controller;
import org.springframework.web.bind.annotation.RequestMapping;
//控制器注解
@Controller
public class HelloController2{
    //返回 ModelAndView 对象
    @RequestMapping(value="/helloController2")
    public ModelAndView handleRequest(HttpServletRequest request,
            HttpServletResponse response)
            throws ServletException, IOException {
        //向 Request 域中放入一条信息，给前端 jsp 用
        request.setAttribute("message", "hello,springmvc");
        //返回 jsp 的路径
        return new ModelAndView("hello");
    }
}
```

这个类和 5.2 节的 HelloController 类有以下两点不同。

① 该类不需要继承其他类，只有@Controller 注解。

② handleRequest 方法前有一个@RequestMapping(value="/helloController2")注解。

（2）修改 springapp-servlet.xml 文件，添加如图 5-9 所示的内容。

```
<?xml version="1.0" encoding="UTF-8"?>
<beans xmlns="http://www.springframework.org/schema/beans"
    xmlns:xsi="http://www.w3.org/2001/XMLSchema-instance"
    xmlns:p="http://www.springframework.org/schema/p"
    xmlns:context="http://www.springframework.org/schema/context"
    xmlns:mvc="http://www.springframework.org/schema/mvc"
    xsi:schemaLocation="http://www.springframework.org/schema/mvc
    http://www.springframework.org/schema/mvc/spring-mvc-4.0.xsd
        http://www.springframework.org/schema/beans
        http://www.springframework.org/schema/beans/spring-beans-3.1.xsd
        http://www.springframework.org/schema/context
        http://www.springframework.org/schema/context/spring-context-3.1.xsd">

<!-- 包扫描器，自动扫描以com.ssmbook2020开头的包 -->
   <context:component-scan base-package="com.ssmbook2020"/>
   <mvc:annotation-driven />

    <!-- 配置一个视图解析器 -->
```

❶ 新增context和mvc的命名空间

❷ 注解包扫描配置

图 5-9　添加对注解的支持

在 MyEclipse 中，可以通过可视化的方式添加 XML 中的命名空间（XML Namespace），如图 5-10 所示。打开 springapp-servlet.xml 文件，在选项卡一栏中由 Source 模式切换到 Namespaces 模式，再勾选 context 和 mvc 选项，对应的 springapp-servlet.xml 文件中就会加上 context 和 mvc 命名空间。

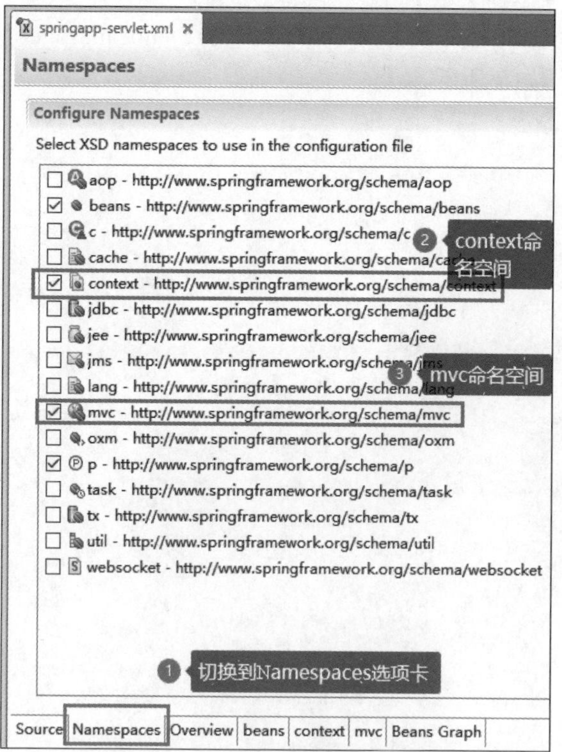

图 5-10　可视化添加命名空间

（3）部署运行。

将项目部署到 Tomcat 服务器，启动 Tomcat 服务器，在浏览器中输入地址 http://localhost:

8080/ssmBook2020_ch5/helloController2.htm。如果配置正确，能看到和 5.2 节同样的 hello.jsp 的页面结果。

在本例中，@RequestMapping(value="/helloController2")中的 value 确定了在用户访问浏览器时该控制器的 URL 地址，注意还要加上.htm 的后缀才能被 Spring MVC 的 Servlet 拦截。

5.5 Spring MVC 注解详解

视频讲解

5.5.1 在类前注解

@RequestMapping 注解除了可以标注在方法上，也可以标注在类上，放于@Controller 之后，标注在类上的作用类似于父路径。例如：

```
@Controller
@RequestMapping("/user")
public class HelloController3{
    //返回 ModelAndView 对象
    @RequestMapping(value="/helloController3")
    public ModelAndView handleRequest(HttpServletRequest request,
            HttpServletResponse response)
            throws ServletException, IOException{
        //向 Request 域中放入一条信息，给前端 jsp 用
        request.setAttribute("message", "hello,springmvc 3");
        //返回 jsp 的路径
        return new ModelAndView("hello");
    }
}
```

此时，访问地址变为 http://localhost:8080/ssmBook2020_ch5/user/helloController3.htm。user 成了 helloController3 的上一级路径，并且无法跳过 user 来访问 helloController3.htm。访问成功的效果如图 5-11 所示。

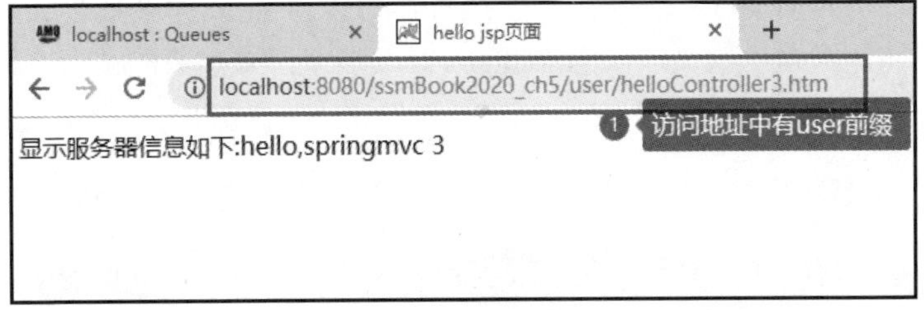

图 5-11 @RequestMapping 标注在类名前面

5.5.2 RequestMapping 注解属性

RequestMapping 注解有 7 个属性，从它的源代码中可以看到 7 个属性的方法名称。

```
public interface RequestMapping extends Annotation {
  //指定映射的名称
    public abstract String name();
  //指定请求路径的地址
    public abstract String[] value();
  //指定请求的方式，是一个RequsetMethod数组，可以配置多个方法
    public abstract RequestMethod[] method();
  //指定参数的类型
    public abstract String[] params();
  //指定请求数据头信息
    public abstract String[] headers();
  //指定数据请求的格式
    public abstract String[] consumes();
  //指定返回的内容类型
    public abstract String[] produces();
}
```

下面主要介绍使用最频繁的 params 和 method 两个属性。

1. params 属性

params 属性表示用户传递的参数名。例如：

```
@Controller
@RequestMapping
public class HelloController4{
    @RequestMapping(value="eg4/select.htm",params="id")
    public String selectById(String id){
        System.out.println("id:"+id);
        return "hello";
    }
}
```

在浏览器中输入地址 http://localhost:8080/ssmBook2020_ch5/eg4/select.htm?id=8 时，id=8 的参数值如果采用纯 Servlet 方式，要使用 request.getParameter()方法才能获得值。而 Spring MVC 可以将其自动注入方法中，在服务器可以得到 id 为 8 的值。这是非常有用的一个功能，可以将用户从一次又一次的 Request 读取多个请求参数的枯燥过程中解放出来。

2. method 属性

method 属性则是指定了只响应指定的请求方式。下面定义一个方法指定 method 为 get。

```
@RequestMapping(value="eg4/test.htm",method=RequestMethod.GET)
    public String test(){
        return "test";
    }
```

在/WEB-INF/jsp/文件夹下新建一个文件 test.jsp，其主要内容如下。

```
<%@ page language="java" import="java.util.*" pageEncoding="UTF-8"%>
```

```
<!DOCTYPE HTML PUBLIC "-//W3C//DTD HTML 4.01 Transitional//EN">
<html>
  <head>
    <title>My JSP 'test.jsp' starting page</title>
  </head>
  <body>
    welocme test.jsp
  </body>
</html>
```

然后在浏览器中输入地址 http://localhost:8080/ssmBook2020_ch5/eg4/test.htm，可以得到如图 5-12 所示的结果页面。

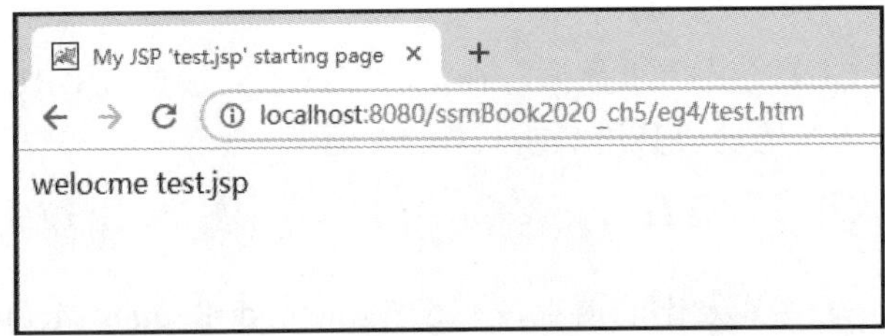

图 5-12　test.jsp 结果页面

如果将其注解改为 method=RequestMethod.POST,再次刷新浏览器，Web 服务器将会报告 405 的错误码：Method Not Allowed，如图 5-13 所示。该提示不允许使用 GET 方式来访问该地址（直接在浏览器中输入地址默认是以 GET 方式访问的）。

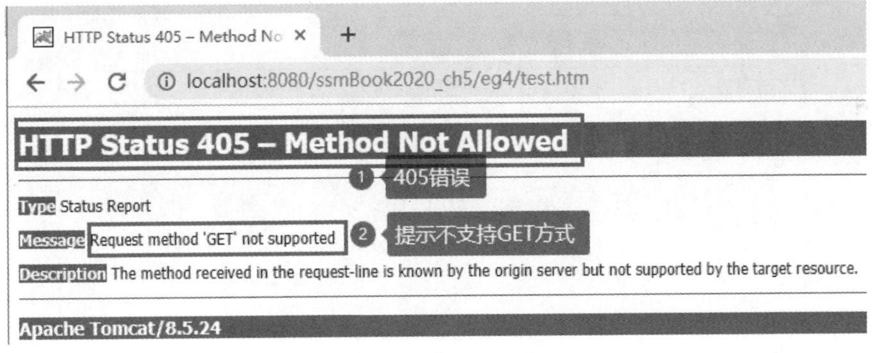

图 5-13　405 错误

为了测试 POST 方式能否正常访问，除了使用 form 表单来设置提交请求的方式之外，还可以使用第三方工具来完成，其中最著名的工具软件是 cURL。

5.5.3　cURL 工具软件

1. cURL 简介

cURL（CommandLine Uniform Resource Locator）是利用 URL 语法在命令行方式下工作的开源文件传输工具。它被广泛应用在 UNIX 和多种 Linux 发行版中，并且有 DOS 和 Windows 32、Windows 64 下的移植版本，用于测试模拟向 Web 服务器发送和接收数据非常方便。

2. cURL 下载

cURL 的下载地址为 https://curl.haxx.se/download.html，下载后解压只有一个文件——curl.exe，将其放到 D 盘的根目录，然后在 Windows 的 DOS 命令窗口中运行它。

cURL 的命令参数非常多，一般用于 RESTful Web Services 测试，涉及以下 4 种参数。

（1）-d/－data <data>：POST 数据内容。

（2）-X/－request <command>：指定请求的方法（使用-d 时自动设为 POST）。

（3）-H/－header <line>：设定 header 信息。

（4）-I/－head：只显示返回的 HTTP 头信息。

3. 发送请求

默认 cURL 使用 GET 方式请求数据，在这种方式下直接通过 URL 传递数据。

在命令行中输入访问地址，其语法格式如下所示。

```
curl http://localhost:8080/ssmBook2020_ch5/eg4/test.htm
```

以上默认是以 GET 方式发出的请求，在设置 method=RequestMethod.*GET* 的情况下可以正确得到返回结果，如图 5-14 所示。

```
@RequestMapping(value="eg4/test.htm",method=RequestMethod.GET)
    public String test(){
        return "test";
    }
```

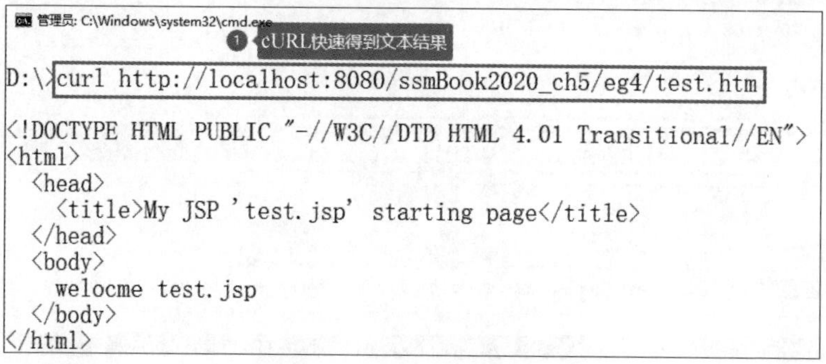

```
管理员: C:\Windows\system32\cmd.exe
① cURL快速得到文本结果

D:\>curl http://localhost:8080/ssmBook2020_ch5/eg4/test.htm

<!DOCTYPE HTML PUBLIC "-//W3C//DTD HTML 4.01 Transitional//EN">
<html>
  <head>
    <title>My JSP 'test.jsp' starting page</title>
  </head>
  <body>
    welocme test.jsp
  </body>
</html>
```

图 5-14　cURL 返回结果

如果使用 POST 请求地址，参数-d 表示以 POST 请求发送数据，其命令如下。

curl -d 0 http://localhost:8080/ssmBook2020_ch5/eg4/test.htm

该命令中的 0 是指无意义地发送数据,以避免将 URL 当成数据而不是地址。服务器返回的是 405 的错误代码。如果将注解的参数改为 method=RequestMethod.POST，再次进行 POST 请求测试，就可以得到正确的页面代码。

cURL 工具对于将来在开发中测试 Web 服务是很方便的，可以大大提高效率。当然除了 cURL，还有一些基于浏览器插件的工具软件，如基于 Chrome 浏览器的 Simple REST Client 和 Postman-REST Client 插件，读者可以根据需要自行下载使用。

5.6 本章小结

本章介绍了如何利用 Spring MVC 框架开发 Java Web 程序。Spring MVC 的核心组件是 DispatcherServlet, 它在项目中响应用户所有的请求，对这些请求进行调度和分发。本章首先通过一个入门案例让大家对 Spring MVC 有了基本的了解，然后讲解了 Spring MVC 的体系结构和原理，最后介绍了使用注解来配置 Spring MVC 控制器的方法，通过注解可以使项目中的配置文件更加简洁。

第 6 章将使用 Spring MVC 重构网上书店案例。

习题 5

1. Spring MVC 对比原生 Servlet 的优点有哪些？
2. 请描述 Spring MVC 框架响应用户请求的过程。
3. ModelAndView 对象的常见使用方法有哪些？

第6章

基于Spring MVC
网上书店重构

本章学习内容

- 使用 Spring MVC 实现会员模块；
- 使用 Spring MVC 实现图书模块；
- 使用 Spring MVC 实现购物车模块；
- 使用 Spring MVC 实现订单模块。

本书第 1 章介绍了 GoodBook 网上书店案例，并实现了显示会员信息的功能。本章使用 Spring MVC 来重构此案例。

6.1 会员模块实现

视频讲解

6.1.1 用户信息显示功能

首先将第 1 章的项目重命名为 bookshopch6，然后加入 Spring MVC 相关的包，可以参考第 5 章的内容。

在项目的主配置文件 web.xml 中加入 Spring MVC 的 DispatcherServlet。然后加入 Spring MVC 的监听器，它会监听 Web 服务的启动，自动读取并加载 Spring 的相关配置文件。详细代码如图 6-1 所示。

```
  <display-name>bookshop</display-name>
  <servlet>
    <servlet-name>springapp</servlet-name>
    <servlet-class>org.springframework.web.servlet.DispatcherServlet</servlet-class>
    <load-on-startup>1</load-on-startup>
  </servlet>
  <servlet-mapping>
    <servlet-name>springapp</servlet-name>
    <url-pattern>*.htm</url-pattern>
  </servlet-mapping>
  <listener>
    <listener-class>org.springframework.web.context.ContextLoaderListener</listener-class>
  </listener>
  <context-param>
    <param-name>contextConfigLocation</param-name>
    <param-value>classpath:applicationContext.xml</param-value>
  </context-param>
    <welcome-file-list>
    <welcome-file>index.html</welcome-file>
     <welcome-file>index.jsp</welcome-file>
  </welcome-file-list>
</web-app>
```

① 定义Spring的DispatcherServlet
② 拦截以htm结尾的请求
③ 监听器类
④ 设置Spring配置文件的位置

图 6-1　项目引入 Spring MVC

将以前的 UserServlet 废弃，改为使用 Spring MVC 的控制器类 UserController，这里使用了注解的方式，代码如下所示。

```
package com.ssmbook2020.web;
//省略导包语句
@Controller
public class UserController {
 @Autowired
 private UserService userService =null;
```

```
    public UserService getUserService(){
        return userService;
    }

    public void setUserService(UserService userService){
        this.userService = userService;
    }

    @RequestMapping("/user/listUser")
    public String listUser(HttpServletRequest request){
        System.out.println("----userController---listuser");
        List<User> userList =userService.getAllUser();
        //将数据放入 Request 域中，以便 jsp 页面可以访问它
        request.setAttribute("userList", userList);
        return "listUser";
    }
}
```

上面的代码中有之前介绍过的@Controller 和@RequestMapping 注解，还有一个新的@Autowired 注解，它用于自动注入对象。那么这个 userService 对象从哪里生成呢？在回答该问题前，先给出一张类之间的依赖关系图。读取用户信息的三层架构的依赖关系如图 6-2 所示。

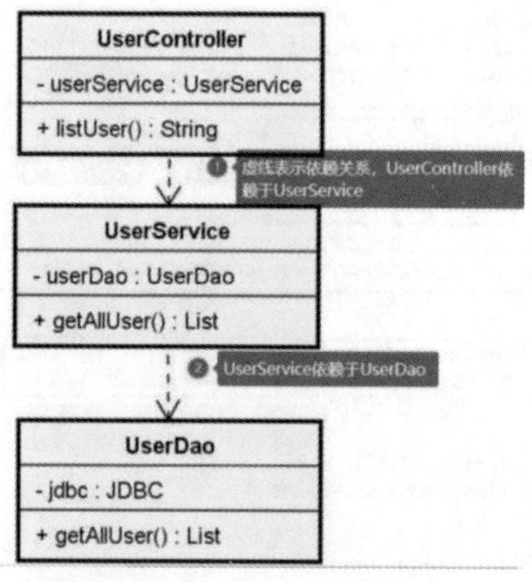

图 6-2　三层架构的依赖关系图

由图 6-2 可知，UserController 依赖于 UserService，而 UserService 又依赖于 UserDao，UserDao 借助于 Jdbc 从数据库中读取数据。

那么 UserServiceImpl 类将会重构成如下的代码，这里只列出了变化的代码。

```
@Service("userService")
public class UserServiceImpl implements UserService {
    @Autowired
    private UserDAO userdao =null;
```

```
    @Override
    public List<User> getAllUser(){

        return userdao.getAllUser();
    }
    public UserDAO getUserdao(){
        return userdao;
    }
    public void setUserdao(UserDAO userdao){
        this.userdao = userdao;
    }
   //省略了部分未实现的方法
}
```

在上述代码中，@Service("userService")定义了一个 id 名为 userService 的 Bean，Spring 将会自动根据名字进行匹配，那么 UserController 类的@Autowired 将会扫描并匹配成功该对象，将 UserServiceImpl 的对象注入。同理，UserServiceImpl 也依赖于 UserDao 对象，那么 UserDao 的实现类 UserDaoImpl 将会重构成如下的代码。

```
@Repository("userdao")
public class UserDaoImpl extends BaseDAOMySQL implements UserDAO {
    @Override
    public List<User> getAllUser(){
        String sql="select * from tbUser";
        List<User> userList =new ArrayList<User>();
        try{
            conn =super.getConnection();
            pstmt = conn.prepareStatement(sql);
            rs = pstmt.executeQuery(sql);
            while (rs.next()){
            String userid = rs.getString("userid"); //
                String username = rs.getString("username");
                String password=""; //不查询密码
                String email = rs.getString("email");
                String phone = rs.getString("phone");
                User user =new User(userid,username,password,email,phone);
                userList.add(user);
            }
        } catch (SQLException e) {
            e.printStackTrace();
        } finally {
            DbUtils.closeQuietly(conn, pstmt, rs);
        }
         return userList;
    }
//省略未实现的代码
}
```

在上述代码中，@Repository("userdao")的注解用来告诉 Spring 这是一个数据库层面的 Bean。 Spring 会根据 3 个类之间的依赖关系，完成自动注入。

现在将此项目部署到 Tomcat 服务器中进行测试，访问地址为 http://localhost:8080/bookshopch6/user/listUser.htm。如果配置正确将会得到如图 6-3 所示的结果。

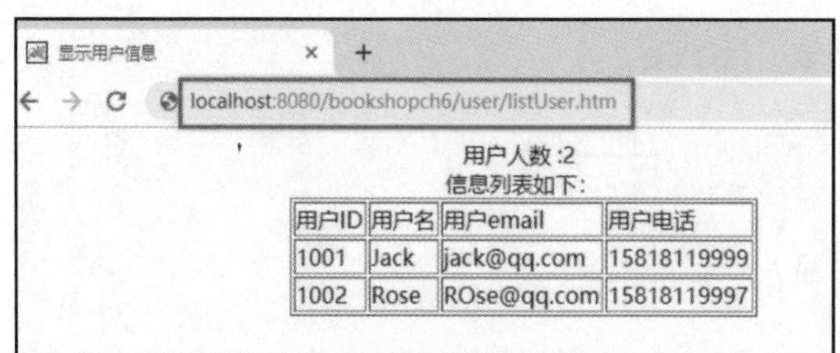

图 6-3　使用注解显示用户列表

6.1.2　会员注册和登录功能

6.1.1 节的用户信息显示功能在整个 Goodbook 网上书店中使用场景是给管理员使用的。在实际系统中，普通用户可以注册和登录，即 1.3.1 节表 1-2 中编号为 1 和 2 的用户故事。现在来实现这两个功能。

（1）在 UserController 控制器中新增注册功能。

在项目中新建一个页面，名为 reg.jsp，其主要内容如下。

```
<%@ page language="java" import="java.util.*" pageEncoding="utf-8"%>
<html>
  <head>
    <title>注册页面</title>
       </head>
    <body>
   <form action="user/regUser.htm" method="post" name="frmreg" onSubmit=
"return validate();">
用户名: <input name="username" type="text" id="username">
密码: <input name="passwd" type="password" id="passwd">
密码确认:<input name="txtconfirm" type="password" id="txtconfirm" >
电话:<input name="phone" type="text" id="phone" >
 E-mail: <input name="email" type="text" id="email" >
<input name="Submit" type="submit" value="确认">
<input type="reset" name="Reset" value="取消">
<a href="login.jsp">已有用户名? 点击这里登录</a>
</form>
 </body>
 </html>
```

Spring MVC 有多种方式可以获取用户注册页面表单中的填写内容，在此介绍其中两种。

① 通过 HttpServletRequest 接收。

② 通过ModelAttribute接收实体对象模型。

通过以下的例子进行测试，User对象的属性都可以获取用户在表单中填写的值，并且repeatPwd也可以获取该值。例如，要获取以上表单中用户输入的用户名。

```
@RequestMapping("/user/regUser")
    public String regUser(@ModelAttribute User user, String repeatPwd,
            HttpServletRequest request){
        boolean flag=false;
        String regResult="";
        int resultNum = userService.saveUser(user);
        if(resultNum>0){
            flag=true;
        }
        if(flag){
            regResult="注册成功";
            request.setAttribute("regResult", regResult);
            return "regSuccess";
        }else{
            regResult="注册失败";
            request.setAttribute("regResult", regResult);
            return "regFail";
        }
    }
```

（2）在业务层类UserServiceImpl中调用UserDaoImpl对象，在UserDaoImpl中使用SpringJDBC提供的JdbcTemplate对象来实现数据库的操作。

整个操作调用流程为UserController→UserService→UserDao→JdbcTemplate。

① 首先从最底层的数据库访问层开始。在Spring配置文件中定义数据源和JdbcTemplate对象，applicationContext.xml代码如下所示。

```
    <?xml version="1.0" encoding="UTF-8"?>
    <beans xmlns="http://www.springframework.org/schema/beans"
        xmlns:xsi="http://www.w3.org/2001/XMLSchema-instance"
        xmlns:p="http://www.springframework.org/schema/p"
        xmlns:context="http://www.springframework.org/schema/context"
        xmlns:mvc="http://www.springframework.org/schema/mvc"
        xsi:schemaLocation="http://www.springframework.org/schema/mvc
http://www.springframework.org/schema/mvc/spring-mvc-3.1.xsd
            http://www.springframework.org/schema/beans
http://www.springframework.org/schema/beans/spring-beans-3.1.xsd
            http://www.springframework.org/schema/context
http://www.springframework.org/schema/context/spring-context-3.1.xsd">
      <!-- 包扫描器，自动扫描以 com.ssmbook2020 开头的包 -->
       <context:component-scan base-package="com.ssmbook2020"/>
      <!-- 定义数据源 -->
    <!-- 使用 apache 的数据源 -->
```

```
    <bean id="dataSource" class="org.apache.commons.dbcp.BasicDataSource"
        destroy-method="close">
        <!-- 配置数据库驱动 -->
        <property name="driverClassName" value="com.mysql.jdbc.Driver" />
        <!-- 数据库连接字符串 -->
        <property name="url" value="jdbc:mysql://localhost:3306/bookdb" />
        <!-- 用户名 -->
        <property name="username" value="root" />
        <!-- 密码 -->
        <property name="password" value="123456" />
        <!--initialSize: 初始连接数 -->
        <property name="initialSize" value="5" />
        <!--maxIdle: 最大空闲连接数 -->
        <property name="maxIdle" value="10" />
        <!--minIdle: 最小空闲连接数 -->
        <property name="minIdle" value="5" />
        <!--maxActive: 最大连接数 -->
        <property name="maxActive" value="30" />
    </bean>


    <bean id="jdbcTemplate" class="org.springframework.jdbc.core.
JdbcTemplate">
        <property name="dataSource" ref="dataSource" />
    </bean>
</beans>
```

② 自定义UserDao的实现类，为了和第1章的UserDaoImpl实现类有所区别，新建类名为UserDaoSpringJDBCImpl.java，代码如下所示。

```
package com.ssmbook2020.dao;
//省略导包语句，具体见源代码
@Repository("userdao")
public class UserDaoSpringJDBCImpl extends BaseDAOMySQL implements UserDAO {

    @Autowired
    private JdbcTemplate jdbcTemplate = null;

    public JdbcTemplate getJdbcTemplate(){
        return jdbcTemplate;
    }

    public void setJdbcTemplate(JdbcTemplate jdbcTemplate){
        this.jdbcTemplate = jdbcTemplate;
    }

    @Override
```

```
        public List<User> getAllUser(){
            String sql = "select * from tbUser";
            return jdbcTemplate.query(sql,new UserRowMapper(),null);

        }

        class UserRowMapper implements RowMapper<User>{
            @Override
            public User mapRow(ResultSet rs, int arg1) throws SQLException{
                int userid = rs.getInt("userid"); //
                String username = rs.getString("username");
                String password = rs.getString("password"); //不查询密码
                String email = rs.getString("email");
                String phone = rs.getString("phone");
                User user = new User(userid, username, password, email, phone);
                return user;
            }
        }

        //返回自动增加的 id 号
        @Override
        public int saveUser(final User user){
            final String sql = "insert into tbUser(username,password,phone,email)
values(?,?,?,?)";
            KeyHolder keyHolder = new GeneratedKeyHolder();

            jdbcTemplate.update(new PreparedStatementCreator() {
                @Override
                public PreparedStatement createPreparedStatement(java.sql.
Connection con) throws SQLException{
                    PreparedStatement ps = con.prepareStatement(sql, Statement.
RETURN_GENERATED_KEYS);
                    ps.setString(1, user.getUsername());
                    ps.setString(2, user.getPasswd());
                    ps.setString(3, user.getPhone());
                    ps.setString(4, user.getEmail());

                    return ps;
                }
            }, keyHolder);
            return keyHolder.getKey().intValue(); //返回自动增加的 id 号;
        }

        @Override
        public int updateUser(final User user){
            final String sql = "update tbUser set username=?,password=?, phone =?,
email=? where userid=?";
```

```
        return jdbcTemplate.update(new PreparedStatementCreator() {
            @Override
            Public PreparedStatement createPreparedStatement(java.sql.
Connection con) throws SQLException {
                PreparedStatement ps = con.prepareStatement(sql);
                ps.setString(1, user.getUsername());
                ps.setString(2, user.getPasswd());
                ps.setString(3, user.getPhone());
                ps.setString(4, user.getEmail());
                ps.setInt(5, user.getUserid());
                return ps;
            }
        });

    }

    @Override
    public User getUserByCondition(final User user){
        String sql = "select * from tbUser where username=? and password=?";

        jdbcTemplate.query(sql, new Object[] { user.getUsername(), user.
getPasswd() }, new RowCallbackHandler() {

            @Override
            public void processRow(ResultSet rs) throws SQLException{
                int userid = rs.getInt("userid"); //

                String email = rs.getString("email");
                String phone = rs.getString("phone");
                user.setUserid(userid);
                user.setEmail(email);
                user.setPhone(phone);

            }
        });
        return user;
    }

    /**
     * 返回对应的用户名密码是否正确
     * @param username
     * @param upass
     * @return
     */
    @Override
    public int checkUser(String username, String upass){
```

```
        String sql = "select count(*) from tbUser where username=? and
password=?";

        return jdbcTemplate.queryForInt(sql, new Object[]{username,upass });

    }

    /**
     * 返回对应的用户名是否存在
     * @param username
     * @return
     */
    @Override
    public int checkUsername(String username){
        String sql = "select count(*) from tbUser where username=?";
        return jdbcTemplate.queryForInt(sql, new Object[] { username });
    }
}
```

③ 在业务层中调用 UserDao 接口中的方法，完成数据库的相关操作，主要有用户注册、用户资料更新、用户登录验证等方法，代码如下。

```
package com.ssmbook2020.service;
//省略导包语句，具体见源代码
@Service("userService")
public class UserServiceImpl implements UserService {

    @Autowired
    private UserDAO userdao =null;

    @Override
    public List<User> getAllUser(){

        return userdao.getAllUser();
    }


    public UserDAO getUserdao(){
        return userdao;
    }

    public void setUserdao(UserDAO userdao){
        this.userdao = userdao;
    }

/**
 * 注册用户时，先检查用户名是否存在。如存在则返回 0，不存在则注册后返回新用户的 ID
```

```
 */
    @Override
    public int saveUser(User user){
        int x= userdao.checkUsername(user.getUsername());
        if(x>0){
            return 0;//用户名存在
        }else{
            return userdao.saveUser(user);
        }

    }

    @Override
    public int updateUser(User user){

        return userdao.updateUser(user);
    }

    @Override
    public User getUserByCondition(User user){
        return userdao.getUserByCondition(user);
    }



   /**
    * 验证用户和密码，用于登录
    */
    @Override
    public User checkLogin(String username, String upass){
    int row=   userdao.checkUser(username, upass);
        if(row>0){
            return userdao.getUserByCondition(new User(username,upass));
        }else{
            return null;
        }

    }

}
```

（3）在项目中引入 JUnit，测试业务层中的方法。

```
package com.ssmbook2020.test;
//省略导包语句，具体可见源代码

public class UserServiceTest {
```

```
        ApplicationContext context = null;

        @Before
        public void init(){
            context = new ClassPathXmlApplicationContext("applicationContext
.xml");
        }

        @Test
        public void testSaveUserService(){
            UserService userservice = (UserService) context.getBean
("userService");
            User u = new User("testuser1006", "123456", "test@qq.com",
"15818111809");
            int newuserid = userservice.saveUser(u);
            System.out.println("新用户 ID:"+newuserid);
        }

        @Test
        public void testUserLogin(){
            UserService userservice = (UserService) context.getBean
("userService");
            User user = userservice.checkLogin("bob", "123456");
            //输入正确的用户和密码，返回对象为非空
            org.junit.Assert.assertNotNull(user);
            User user2 = userservice.checkLogin("bob", "1234567");
            //输入错误的用户和密码，返回对象为非空
            org.junit.Assert.assertNull(user2);
        }
    }
```

在测试类中主要测试新用户注册和登录功能，以上方法均顺利通过测试。

（4）从界面到后端对功能进行完整测试。

测试登录界面 login.jsp，其代码如下。

```
    <%@ page language="java" import="java.util.*" pageEncoding="utf-8"%>
    <html>
      <head>
        <title></title>
        <style type="text/css">
         //省略样式，具体见源代码
        </style>
        <script language="JavaScript" type="text/JavaScript">

        function validate()
        {
         //表单验证,省略
```

```
        return true;
        }
        </script>
        </head>

        <body>
          <div align="center">
            <p><img src="img/goodbooklogo.png" width=280 height="140"/><br/>
</p>
            <p><font color="#3300FF"><b>欢迎来到GoodBook网上书店</b></font></p>
          </div>
          <form action="user/loginUser.htm" method="post" name="frmreg"
onSubmit="return validate()">
            <table width="500"  border="0" align="center">
              <tr align="center" valign="top">
                <td colspan="2"> <strong>请输入用户名密码</strong></td>
              </tr>
              <tr>
                <td width="25%"><span class="style23 style36 style37"><strong>
用户名：</strong></span></td>
                <td width="75%">
                  <input name="username" type="text" id="username" size="30"
maxlength="35">
                  </td>
                </tr>
                <tr>
                  <td><span class="style23 style36 style37"><strong>密码:</strong>
</span></td>
                  <td>
                    <input name="passwd" type="password" id="passwd" size="12"
maxlength="10">
                    </td>
                  </tr>
                                  <tr>
                                    <td> </td>
                                    <td>
                                     <input name="Submit" type="submit" value="确认">
                                      <input type="reset" name="Reset" value="取消">
                                        </td>
                                      </tr>
                                   </tr>
                                    <td> </td>
                                    <td>
                                      <a href="reg.jsp">没有用户名？点击这里注册</a>
                              </td>
                         </tr>
                  </table>
```

```
        </form>
    </body>
  </html>
```

部署运行后的界面如图 6-4 所示。

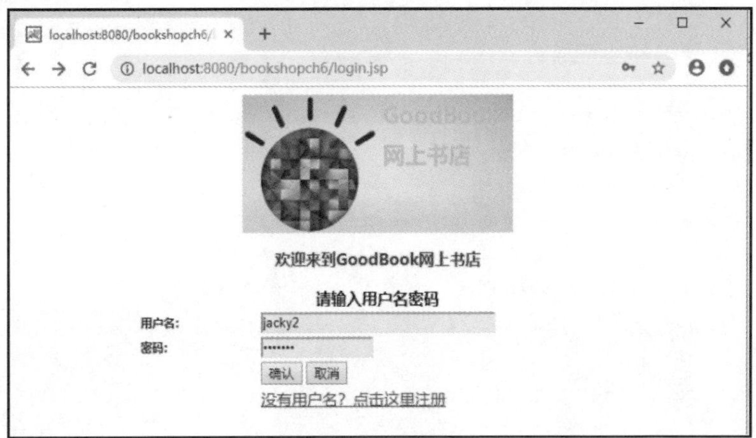

图 6-4　登录界面

在注册功能中，新增 reg.jsp 页面、注册成功界面 regSuccess.jsp 和注册失败界面 regFail.jsp。

（5）在 UserController 控制器中新增登录和注册的方法，其代码如下。

```
package com.ssmbook2020.web;
//省略导包语句，具体见源代码
import com.ssmbook2020.bean.User;
import com.ssmbook2020.service.UserService;
@Controller
public class UserController {

    @Autowired
    private UserService userService = null;

    public UserService getUserService(){
        return userService;
    }

    public void setUserService(UserService userService){
        this.userService = userService;
    }

    @RequestMapping("/user/listUser.htm")
    public String listUser(HttpServletRequest request){
        System.out.println("----userController---listuser");
        List<User> userList = userService.getAllUser();
        //将数据放入 request 域中，以便 jsp 页面可以访问它
```

```
            request.setAttribute("userList", userList);
            return "listUser";
        }

        @RequestMapping("/user/regUser.htm")
        public String regUser(@ModelAttribute User user, String repeatPwd,
HttpServletRequest request){
            String regResult = "";
            int resultNum = userService.saveUser(user);
            if (resultNum > 0){
                regResult = "regSuccess";
                request.setAttribute("regResult", "注册成功");
            } else {
                regResult = "regFail";
                request.setAttribute("regResult", "注册失败");
            }
             return regResult;
        }

        @RequestMapping("/user/loginUser.htm")
        public String checkLogin( String username,  String passwd,
HttpServletRequest request){
            User user = userService.checkLogin(username, passwd);
            String result = "";
            if (user != null){
                request.getSession().setAttribute("loginUser", user);
                result = "loginSuccess";
            } else {
                request.setAttribute("loginFail", "用户名或密码不正确");

                result = "loginFail";
            }
            return result;
        }

    }
```

6.2 图书模块实现

网上书店的核心是图书销售，包括待销售图书的显示和用户网上购买图书。下面要实现图书的显示和销售模块，对应 1.3.1 节表 1-2 中用户故事编号 3 和 4 的功能。

图书模块的实现步骤如下。

（1）定义图书类。

```
public class BookBean implements Serializable {
    private String bookId="";
    private String ISBN="";      //ISBN编号
    private String bookname=""; //书名
    private String author="";    //作者名字
    private String imageFile="";//封面图像
    private int editionNumber=1;//版本
    private String publisher="";//出版商
    private double price=0.0d;  //价格
    private double discount=1;  //折扣
    private DecimalFormat df = new DecimalFormat("0.00");
    //省略构造函数和所有属性的get/set 函数
}
```

（2）定义图书数据访问层接口类。

```
public interface BookDAO {
    //显示所有图书
    public List<BookBean> getAllBooks();
    //根据书名或作者等信息查询图书
    public List<BookBean> getBooksByCondition(String bookId,String booksName,
String author,String isbn);
     //根据图书id查询图书
    public BookBean getBookById(String bookId);
    //查询商品种类数量
    public int getGoodsCount();
    //新增图书
    public int addBook(BookBean bookBean);
    //根据图书id修改图书信息
    public int updateBook(BookBean bookBean ,String bookId);
}
```

（3）定义BookDaoImpl类，它实现了BookDAO接口，使用JDBC完成对数据库的读写操作。

```
@Repository("bookdao")
public class BookDaoImpl extends BaseDAOMySQL implements BookDAO {
    @Autowired
    private JdbcTemplate jdbcTemplate = null;
    @Override
    public List<BookBean> getAllBooks(){
        String sql="select * from tbbook order by bookid";
        RowMapper<BookBean> rowMapper=new BeanPropertyRowMapper<BookBean>
(BookBean.class);
        return jdbcTemplate.query(sql, rowMapper);

    }
```

```
    @Override
    public BookBean getBookById(String bookId){
        String sql="select * from tbbook  where 1=1 and bookId=? ";
        Object[] params={bookId};
        RowMapper<BookBean> rowMapper=new BeanPropertyRowMapper<BookBean>
(BookBean.class);
        return jdbcTemplate.queryForObject(sql,params, rowMapper);
       //省略其他方法
    }
```

注意，这个类名前面有@Repository 的注解。@Repository("bookdao")注解相当于在 Spring 中定义了一个名为 bookdao 的 Java Bean。

（4）定义服务层接口 BooService 类，然后再定义 BookServiceImpl 类实现 BookService 接口，其代码如下。

```
public interface BookService {
      //得到所有的用户信息
        public List<BookBean> getAllBook();

         //根据 ID 得到图书
        public BookBean  getBookById(String bookId);

         //根据 ISBN 得到图书
        public BookBean  getBookByISBN(String bookIsbn);

        //根据条件，查找某个用户
        public List<BookBean> getBookByCondition(BookBean book);
}

 @Service("bookService")
public class BookServiceImpl implements BookService {

    @Autowired
    BookDAO bookdao =null;

    public BookDAO getBookdao(){
        return bookdao;
    }

    public void setBookdao(BookDAO bookdao){
        this.bookdao = bookdao;
    }

    @Override
    public List<BookBean> getAllBook(){
        return bookdao.getAllBooks();
    }
```

```
    @Override
    public List<BookBean> getBookByCondition(BookBean book){
        System.out.println(book);
        return
        bookdao.getBooksByCondition(book.getBookId(),book.getBookname(),
        book.getAuthor(), book.getISBN());
    }

  //根据图书 id 查询图书
    @Override
    public BookBean getBookById(String bookId){
        return bookdao.getBookById(bookId);
    }
    //根据 ISBN 查询图书未实现
}
```

（5）定义基于 SpringMVC 的控制器 BookController 类，它用来替代传统 Servlet 所起的作用。

```
@Controller
public class BookController {
    @Autowired
    BookService bookService =null;

    public void setBookService(BookService bookService){
        this.bookService = bookService;
    }
    @RequestMapping("/book/listBook.htm")
    public String listBook(HttpServletRequest request){
        List<BookBean> bookList = bookService.getAllBook();
        //将数据放入 Request 域中，以便 jsp 页面可以访问它
        request.setAttribute("bookList", bookList);
        return "listBook";
    }
    @RequestMapping("/book/search.htm")
    public String searchBook(HttpServletRequest request){

        String bookName=request.getParameter("bookname");
        String author=request.getParameter("author");
        String isbn=request.getParameter("isbn");
        BookBean book =new BookBean(bookName,author,isbn);

        List<BookBean> bookList = bookService.getBookByCondition(book);
        //将数据放入 Request 域中，以便 jsp 页面可以访问它
```

```
        request.setAttribute("bookList", bookList);
        return "listBook";
    }

}
```

在以上代码中，BookController 依赖于 BookService 类，而 BooService 类依赖于 BookDao 接口。

```
@RequestMapping("/book/listBook.htm")
    public String listBook(HttpServletRequest request){
        List<BookBean> bookList = bookService.getAllBook();
        //将数据放入 Request 域中，以便 jsp 页面可以访问它
        request.setAttribute("bookList", bookList);
        return "listBook";
    }
```

listBook 方法最后的返回值为“listBook”字符串，根据 SpringMVC 的视图映射规则，对应项目的地址为/WEB-INF/jsp/listBook.jsp。

（6）在 listBook.jsp 中，使用 JSTL 和 EL 表达式，完成显示图书的相关信息。在页面中引入 JSTL 的标签库，其代码如下。

```
<!-- 导入 jstl 的标准核心库 -->
<%@taglib uri="http://java.sun.com/jsp/jstl/core" prefix="c" %>
<!-- 导入 jstl 的函数库 -->
<%@taglib uri="http://java.sun.com/jsp/jstl/functions" prefix="fn" %>
<!-- 导入 jstl 的格式化库 -->
<%@taglib uri="http://java.sun.com/jsp/jstl/fmt" prefix="fmt" %>
```

图书显示的主要代码如下。

```
<div class="div4">
    找到的结果数为 :${fn:length(requestScope.bookList) } <br/>
   </div>
    <div>
   <table border="" style="margin:auto" >
     <thead>
      <tr>
        <td>图片</td> <td>图书名字</td>
          <td>作者</td> <td>原价（￥）</td>
          <td>折扣</td> <td>折后价（￥）</td>
          <td>出版社</td>  <td>ISBN 号</td>
           <td>操作</td>
      </tr>
     </thead>
       <tbody>
       <c:forEach items="${requestScope.bookList }" var="book">
          <tr>
```

```
        <td><img src="../img/${book.imageFile}" width="144" height="96" /> </td>
                <td>${book.bookname}</td>
                <td>${book.author}</td>

                <td><fmt:formatNumber value="${book.price}" type="currency"
pattern="￥.00" /></td>
                <td><fmt:formatNumber value="${book.discount}" type="percent"
pattern=".00%" /></td>
                 <td><fmt:formatNumber value="${book.price * book.discount}"
type="currency" pattern="￥.00" /></td>
                <td>${book.publisher}</td>
                <td>${book.ISBN }</td>
                  <td>
                    <div id="buttonDiv${book.bookId }"><a id="cart" href="#"
onclick="addCart(${book.bookId })">加入购物车</a></div>
                   <div style="display:none" class="divresult" id="div${book.
bookId }">已加入购物车</div>

                  </td>
            </tr>
         </c:forEach>
        </tbody>
      </table>
      </div>
```

重新部署项目到 Tomcat 服务器下，并启动 Tomcat 服务器。在浏览器中输入地址 http://localhost:8080/bookshopch6/book/listBook.htm，显示结果如图 6-5 所示。

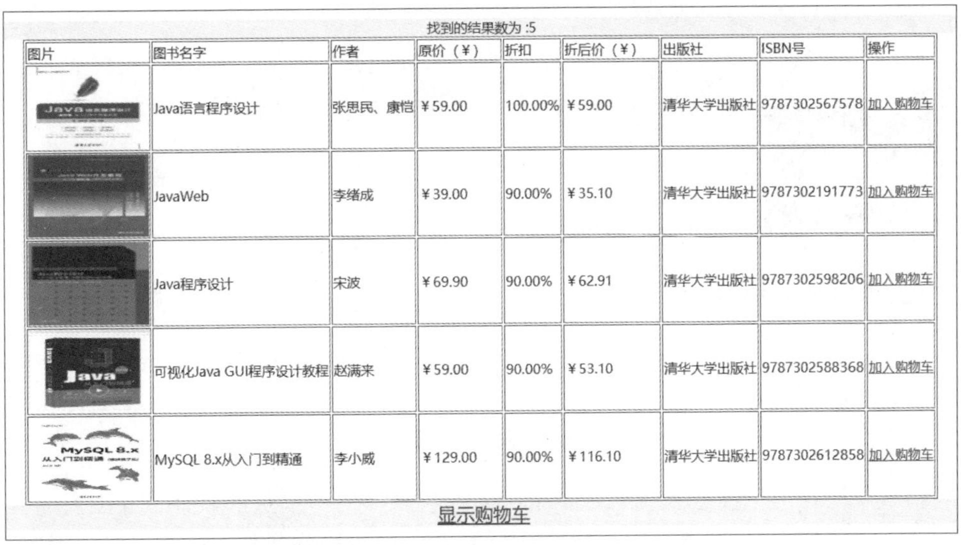

找到的结果数为 :5

图片	图书名字	作者	原价（¥）	折扣	折后价（¥）	出版社	ISBN号	操作
	Java语言程序设计	张思民、康恺	¥59.00	100.00%	¥59.00	清华大学出版社	9787302567578	加入购物车
	JavaWeb	李绪成	¥39.00	90.00%	¥35.10	清华大学出版社	9787302191773	加入购物车
	Java程序设计	宋波	¥69.90	90.00%	¥62.91	清华大学出版社	9787302598206	加入购物车
	可视化Java GUI程序设计教程	赵满来	¥59.00	90.00%	¥53.10	清华大学出版社	9787302588368	加入购物车
	MySQL 8.x从入门到精通	李小威	¥129.00	90.00%	¥116.10	清华大学出版社	9787302612858	加入购物车

显示购物车

图 6-5　图书列表显示页面

6.3 购物车模块实现

在图书列表显示页面中，如果要将喜欢的书放入购物车，该如何操作呢？这里采用超链接的方式，直接将想要的书放入购物车中。为了有更好的体验，该操作采用 AJAX 技术，无须刷新整个页面，就可以将选中的图书放入购物车。将商品放入购物车如图 6-6 所示，显示购物车信息如图 6-7 所示。

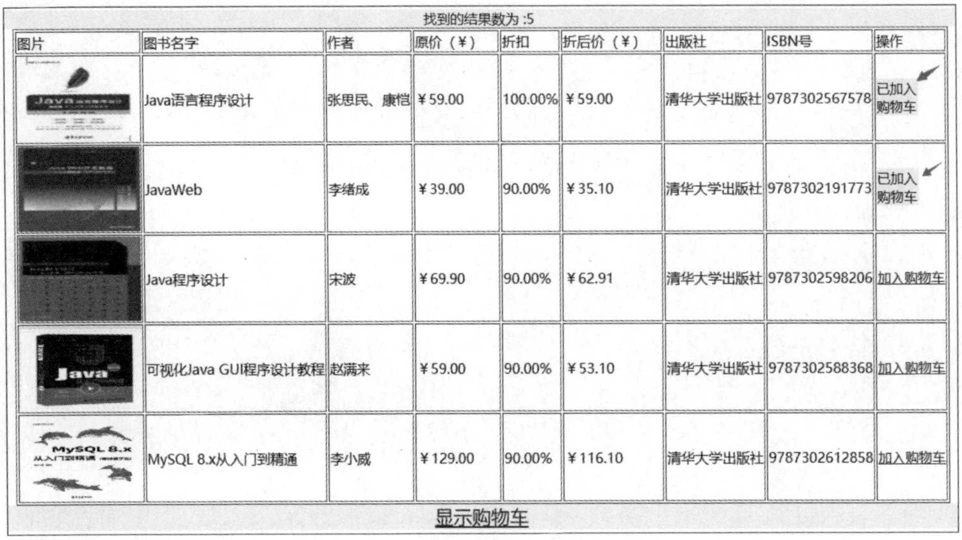

找到的结果数为 :5

图片	图书名字	作者	原价（¥）	折扣	折后价（¥）	出版社	ISBN号	操作
	Java语言程序设计	张思民、康恺	¥59.00	100.00%	¥59.00	清华大学出版社	9787302567578	已加入购物车
	JavaWeb	李绪成	¥39.00	90.00%	¥35.10	清华大学出版社	9787302191773	已加入购物车
	Java程序设计	宋波	¥69.90	90.00%	¥62.91	清华大学出版社	9787302598206	加入购物车
	可视化Java GUI程序设计教程	赵满来	¥59.00	90.00%	¥53.10	清华大学出版社	9787302588368	加入购物车
	MySQL 8.x从入门到精通	李小威	¥129.00	90.00%	¥116.10	清华大学出版社	9787302612858	加入购物车

显示购物车

图 6-6　将商品放入购物车

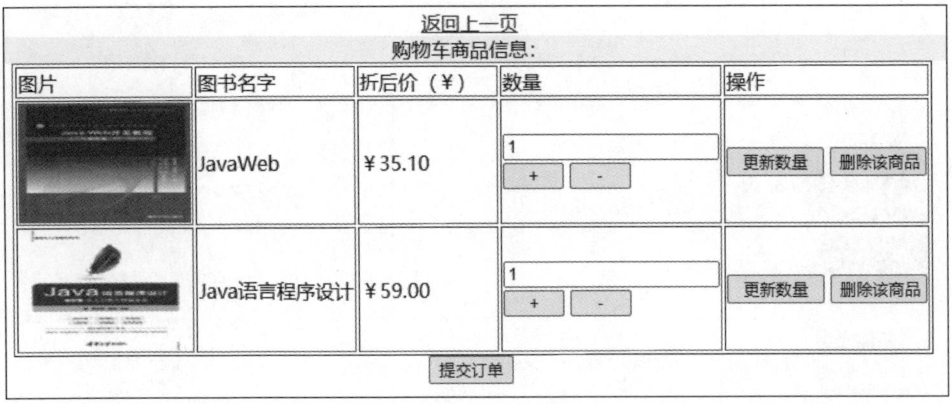

返回上一页

购物车商品信息:

图片	图书名字	折后价（¥）	数量	操作
	JavaWeb	¥35.10	1 + -	更新数量 删除该商品
	Java语言程序设计	¥59.00	1 + -	更新数量 删除该商品

提交订单

图 6-7　显示购物车信息

购物车模块在电子商务系统中是一个常见的模块，这个模块的具体设计因人而异。复杂版设计能永久保存用户的购物车信息，但需要在后端程序和数据库中建立对应表的支持。而简化版设计的数据只在当前登录状态下有效，本次会话过期后，购物车就被清空了，数据没有被保存在数据库中。本演示项目采用简化设计，但是复杂版和简化版并不冲突，在简化版功能的基础上，增加购物车持久化到数据库中即可实现复杂版。

（1）新建一个购物车列表项实体类，该类包含图书对象和数量，主要目的是方便图 6-7 中每项图书商品的显示。

```
public class CartItemBean implements Serializable {

    private BookBean book;

    private int quantity;
   //省略 get/set 方法 构造函数等
}
```

（2）在 BookController 中新增加入图书到购物车的方法 addBookToCart。由于采用 AJAX 请求，该方法返回值为字符串。

```
@RequestMapping(value="/book/addCart.htm")
 @ResponseBody
public String addBookToCart(HttpServletRequest request,
        @RequestParam("bookId")String bookId) {
    BookBean book= bookService.getBookById(bookId);
    HttpSession session= request.getSession();
    @SuppressWarnings("unchecked")
    Map<String,CartItemBean> cartMap=
                (Map)session.getAttribute("cartMap");

    if(cartMap==null){
        cartMap =new HashMap<String,CartItemBean>();
     }
    CartItemBean cartItem=(CartItemBean)cartMap.get("bookId");
    if(cartItem==null){
         cartItem =new CartItemBean(book,1);
    }else{
        cartItem.setQuantity(cartItem.getQuantity()+1);
    }
    cartMap.put(bookId, cartItem);
    session.setAttribute("cartMap", cartMap);
    return "success";
}
```

（3）页面中使用 JQuery 的 AJAX 发送将某本书加入到购物车的请求，接收到返回值后判断是否加入购物车。

```
    <script type="text/javascript" src="<%=request.getContextPath() %>/js/
jquery-3.2.1.min.js"></script>
    <script type="text/javascript">
    function addCart(bookId){
       var divBook="#div"+bookId
       var buttonDiv="#buttonDiv"+bookId
       var params="bookId="+bookId
```

```
    $.post("addCart.htm",params,function(operateSuccess){
        if(operateSuccess=="success"){
            $(divBook).show();
          $(buttonDiv).hide();
        }else{
          alert("加入购物车失败，请稍候再试");
        }
  });
  return false;
}
```

在 listBook.jsp 页面中，发送加入购物车请求的超链接如下。

```
<a id="cart" href="#" onclick="addCart(${book.bookId })">加入购物车</a>
```

（4）显示购物车内容。

购物车显示页面为 listCart.jsp，它的主要思路是在显示 BookController 类的 addBookToCart 方法中放入 Session 域的信息。获取购物车数据的关键代码如下。

```
    Map<String,CartItemBean> cartMap=(Map)session.getAttribute("cartMap");
if(cartMap==null){
        cartMap =new HashMap<String,CartItemBean>();
    }
    CartItemBean cartItem=(CartItemBean)cartMap.get("bookId");
    if(cartItem==null){
          cartItem =new CartItemBean(book,1);
    }else{
         cartItem.setQuantity(cartItem.getQuantity()+1);
    }
    cartMap.put(bookId, cartItem);
    session.setAttribute("cartMap", cartMap);
```

可以看出该对象被设计成 Map 接口对象。购物车数据结构如表 6-1 所示。

表 6-1　购物车数据结构

键	值
图书 id	CartItem 对象
示例键 bookid:1001	示例值 CartItem{bookid：1001,数量：2}
示例键 bookid:1002	示例值 CartItem{bookid：1002,数量：1}

（5）更改购物车中的商品数量。

为了方便用户更改购物车中的商品数量，可以增加两个按钮，分别用于增加和减少商品的数量。在 listCart.jsp 中使用 JavaScript 代码完成更新操作，所用函数定义如下。

```
/** 更新购物车中的商品数量 */
function updateItem(bookNum,bookId){
 var params="bookId="+bookId+"&bookNum="+bookNum;
```

```
$.post("updateCart.htm",params,function(operateSuccess){
      if(operateSuccess=="updateSuccess"){
        //alert("已更新购物车");
        var newValue=parseFloat($("#quantityTxt"+bookId).val())+bookNum
        if(newValue<0){
        newValue=0;
        }
         $("#quantityTxt"+bookId).val(newValue);
      }else{
         alert("更新购物车失败，请稍后再试");
      }
})；
 return false; }
```

该方法向后端服务器发送请求的 URL 地址为 http://localhost:8080/bookshopch6/book/updateCart.htm。后端控制器 BookController 增加对应方法响应该请求，返回结果为字符串，代码如下。

```
/* 更新购物车中的图书数量
 * @param request
 * @return reponseBody 表示返回字符串给 jquery ajax
 */
 @RequestMapping(value="/book/updateCart.htm")
 @ResponseBody
public String updateCart(HttpServletRequest request,
        @RequestParam("bookId")String bookId,int bookNum) {
    BookBean book= bookService.getBookById(bookId);
    HttpSession session= request.getSession();
    Map<String,CartItemBean> cartMap=(Map)session.getAttribute ("cartMap");
    if(cartMap==null){
        cartMap =new HashMap<String,CartItemBean>();
    }
    CartItemBean cartItem=(CartItemBean)cartMap.get(bookId);
    if(cartItem==null){
        cartItem =new CartItemBean(book,1);
    }else{
        cartItem.setQuantity(cartItem.getQuantity()+bookNum);
        if(cartItem.getQuantity()<0){
              cartItem.setQuantity(0);
        }
    }
    cartMap.put(bookId, cartItem);
    session.setAttribute("cartMap", cartMap);
    return "updateSuccess";
}
```

（6）删除购物车中的某件商品。

为了方便用户删除放入购物车的商品，在 listCart.jsp 页面中增加一个删除按钮，其代码如下。

```
<input type="button" value="删除该商品" style="width:80"
onclick="delItem(this,${m.value.book.bookId})"/>
```

该按钮调用当前页面的 Javascript 方法 delItem()，该方法定义的代码如下。

```
/* 删除购物车中的指定商品  */
function delItem(obj,bookId){
    if (confirm("确认删除?")){
     var params="bookId="+bookId;
    $.post("delCart.htm",params,function(operateSuccess){
         alert(operateSuccess);
    if(operateSuccess=="delSuccess"){
            alert("已删除购物车中该商品");
            delRow(obj);//删除表格当前行
           }else{
             alert("更新购物车失败，请稍后再试");
           }
      });
  }
     return false;
}

 //删除表格当前行
function delRow(obj) {
       $(obj).parent().parent().remove();
     }
```

该方法向后端服务器发送请求的 URL 地址为 http://localhost:8080/bookshopch6/book/delCart.htm。后端服务器类 BookController 中增加新方法 delItemCart()响应该请求，代码如下。

```
    @RequestMapping(value="/book/delCart.htm")
        @ResponseBody
       public String delItemCart(HttpServletRequest request,
               @RequestParam("bookId")String bookId) {
           HttpSession session= request.getSession();
           Map<String,CartItemBean>cartMap=(Map)session.getAttribute
("cartMap");
           if(cartMap==null){
               cartMap =new HashMap<String,CartItemBean>();
           }
           cartMap.remove(bookId);
         session.setAttribute("cartMap", cartMap);
           return "delSuccess";
       }
```

（7）提交订单。

用户选购完成后，最后一步就是提交该订单，同时清空购物车。订单模块比较复杂，将在下一节详解介绍。

6.4 订单模块实现

用户最终要提交订单时，需要在订单确认页面填写收件人相关信息。经过分析发现，在第 1 章设计的 E-R 图中，关于订单的部分过于简单。在系统分析中有一个自顶向下、逐步求精的过程。重新设计之后的订单模块分为订单表和订单明细表两部分，它们的实体关系图如图 6-8 所示。

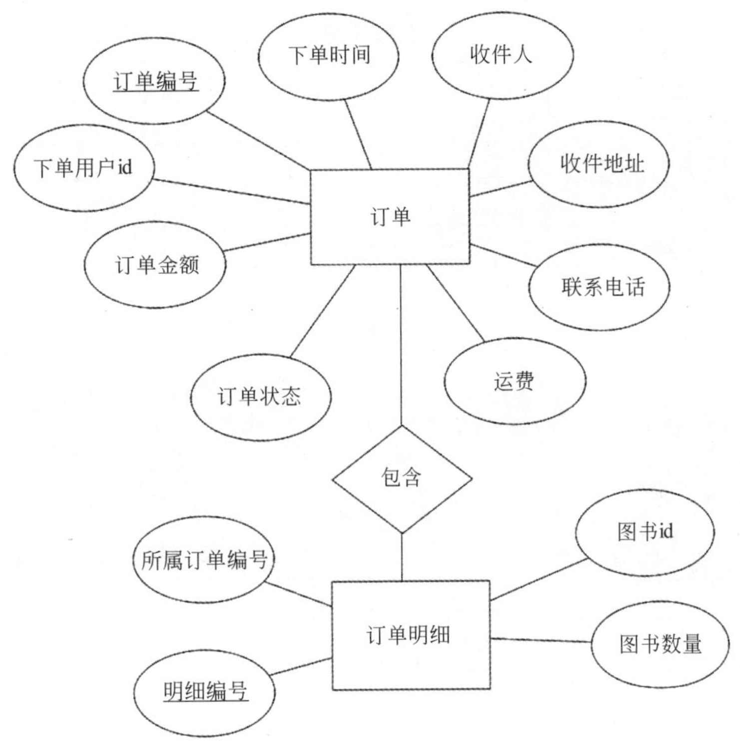

图 6-8 订单表和订单明细表 E-R 图

将数据库逻辑模型转换成数据库的表，订单表如表 6-2 所示，订单明细表如表 6-3 所示。

订单模块的实现步骤如下。

（1）分别完成订单表和订单明细表。

订单表的数据库代码如下。

表 6-2 订单表

字段名	类 型	长 度	小数位数	是否为空	是否主键
orderid	bigint	0	0	否	主键
userid	int	0	0	否	
totalFee	double	10	2	否	
status	int	0	0	是	
orderTime	timestamp	0	0	否	
receiver	varchar	20	0	否	
address	varchar	200	0	否	
phone	varchar	30	0	否	
shipFee	double	0	0	否	
memo	varchar	50	0	否	

表 6-3 订单明细表

字段名	类 型	长 度	小数位数	是否为空	是否主键
orderItemId	int	0	0	否	主键
orderid	bigint	0	0	否	
bookid	int	0	0	否	
bookname	varchar	255	0	否	
disCountPrice	double	10	2	否	
count	int	0	0	否	

```
CREATE TABLE 'tborder' (
  'orderid' bigint unsigned NOT NULL AUTO_INCREMENT,
  'userid' int NOT NULL,
  'totalFee' double(10,2) NOT NULL,
  'status' int DEFAULT NULL,
  'orderTime' timestamp NOT NULL ON UPDATE CURRENT_TIMESTAMP,
  'receiver' varchar(20) CHARACTER SET utf8 COLLATE utf8_general_ci NOT NULL
COMMENT '收货人',
  'address' varchar(200) CHARACTER SET utf8 COLLATE utf8_general_ci NOT NULL,
  'phone' varchar(30) CHARACTER SET utf8 COLLATE utf8_general_ci NOT NULL,
  'shipFee' double NOT NULL,
  'memo' varchar(50) CHARACTER SET utf8 COLLATE utf8_general_ci NOT NULL,
  PRIMARY KEY ('orderid')
) ENGINE=InnoDB AUTO_INCREMENT=100000 DEFAULT CHARSET=utf8;
```

订单明细表的数据库代码如下。

```
CREATE TABLE 'tborderitem' (
  'orderItemId' int unsigned NOT NULL AUTO_INCREMENT,
  'orderid' bigint NOT NULL,
  'bookid' int NOT NULL,
  'bookname' varchar(255) CHARACTER SET utf8 COLLATE utf8_general_ci NOT
```

```
NULL,
    'disCountPrice' double(10,2) NOT NULL,
    'count' int unsigned NOT NULL,
    PRIMARY KEY ('orderItemId')
  ) ENGINE=MyISAM AUTO_INCREMENT=10 DEFAULT CHARSET=utf8;
```

（2）将数据库表和 Java 面向对象类进行映射。定义订单类 Order.java 的代码如下。

```
public class Order {
    private String orderNo;       //订单编号
    private int orderUserId;      //下订单 ren 的用户 id

    private double totalFee;      //订单总金额
    private int status;           //0 刚下单 1 已确认 2 已发货 3 已收货 4 已取消
    private String orderTime;
    private String receiver;      //收货人
    private String address;
    private String phone;
    private double shipFee=0d;    //运费,默认为 0 元
    private String memo;          //备注

    private List<CartItemBean> list = new ArrayList<CartItemBean>();
//省略 getter/setter 方法
//省略构造函数
}
```

（3）在数据访问层完成对订单表的操作，主要方法有保存订单、根据订单编号查询订单和根据用户编号查询订单等。

```
public interface OrderDAO {
       //添加订单，返回订单号
        public String addOrder(Order order);
        //更新订单
        public int updateOrder(Order order);
        //根据订单编号，查询一个订单
        public Order queryOrder(String orderNo);
        //根据用户编号，查询多个订单
        public List<Order> queryOrderByUserId(String userid);

        //添加到订单明细表
        public void addOrderDetail(String orderNo,CartItemBean cartItemBean);
}
```

（4）实现 OrderDAO 接口的实现类 OrderDaoImpl 中的方法。

```
@Repository("orderdao")
public class OrderDaoImpl extends BaseDAOMySQL implements OrderDAO {
```

```
        @Autowired
        private JdbcTemplate jdbcTemplate = null;

        public JdbcTemplate getJdbcTemplate(){
            return jdbcTemplate;
        }

        public void setJdbcTemplate(JdbcTemplate jdbcTemplate){
            this.jdbcTemplate = jdbcTemplate;
        }


        /**
         * 根据用户 id，查询订单
         */
        @Override
        public List<Order> queryOrderByUserId(String orderUserId){
            String sql = "select * from tbOrder where userid=? ";
            return jdbcTemplate.query(sql,new OrderRowMapper(),new Object[]
{orderUserId});
        }

        @Override
        public Order queryOrder(String orderNo){
            String sql = "select * from tbOrder where orderid = ? ";
            return jdbcTemplate.queryForObject(sql,new OrderRowMapper(),new
Object[]{orderNo});
        }


        class OrderRowMapper implements RowMapper<Order>{
            @Override
            public Order mapRow(ResultSet rs, int arg1) throws SQLException{
                String orderNo = rs.getString("orderid"); //
                int userid=rs.getInt("userId");
                double totalFee = rs.getDouble("totalFee");
                int status = rs.getInt("status");
                String orderTime = rs.getString("orderTime");
                String addressee = rs.getString("receiver");//收货人
                String address = rs.getString("address");//收货人地址
                String phone = rs.getString("phone");
                double shipFee  = rs.getDouble("shipFee");
                String memo =rs.getString("memo");//备注
                Order order = new Order(orderNo, userid,totalFee,status,orderTime,
addressee, address, phone,shipFee,memo);
                return order;
            }
```

```
    }

    //返回自动增加的 id 号
    @Override
    public String addOrder(final Order order){
        final String sql = "insert into tbOrder"
                + "(orderid,userId,totalFee,status,orderTime,receiver,
address, phone,shipFee,memo)"
                + "values(default,?,?,?,?,?,?,?,?,?)";
        KeyHolder keyHolder = new GeneratedKeyHolder();

        jdbcTemplate.update(new PreparedStatementCreator() {
            @Override
            public PreparedStatement createPreparedStatement(java.sql.
Connection con) throws SQLException{
                PreparedStatement ps = con.prepareStatement(sql,
Statement.RETURN_GENERATED_KEYS);
                ps.setInt(1, order.getOrderUserId());
                ps.setDouble(2, order.getTotalFee());
                ps.setInt(3, order.getStatus());
                ps.setString(4, order.getOrderTime());
                ps.setString(5, order.getReceiver());
                ps.setString(6,order.getAddress() );
                ps.setString(7, order.getPhone());
                ps.setDouble(8, order.getShipFee());
                ps.setString(9, order.getMemo());
                return ps;
            }
        },keyHolder);
        return keyHolder.getKey()+""; //返回自动增加的 id 号

    }

    @Override
    public int updateOrder(Order order){
        //TODO Auto-generated method stub
        return 0;
    }

    @Override
    public void addOrderDetail(final String orderNo,final CartItemBean
cartItemBean){
        final BookBean bookBean=cartItemBean.getBook();
        final String sql = "insert into tbOrderItem"
                +"(orderItemId,orderid,bookid,bookName,disCountPrice,count)"
                + "values(default,?,?,?,?,?)";
```

```
        jdbcTemplate.update(new PreparedStatementCreator() {
            @Override
            public PreparedStatement createPreparedStatement(java.sql.
Connection con) throws SQLException{
                PreparedStatement ps = con.prepareStatement(sql);
                ps.setString(1, orderNo);
                ps.setString(2, bookBean.getBookId());
                ps.setString(3, bookBean.getBookname());
                ps.setDouble(4, bookBean.getPrice()*bookBean.getDiscount());
                ps.setInt(5, cartItemBean.getQuantity());
                return ps;
            }
        });
    }
}
```

这里可以使用JUnit完成订单类数据库操作方法的单元测试，参见代码OrderDaoTest.java。

（5）定义订单业务层接口OrderService.java和实现类OrderServiceImpl.java。

```
public interface OrderService {
    //添加订单，返回订单号
        public String addOrder(Order order);
        //更新订单
        public int updateOrder(Order order);
        public Order queryOrder(String orderNo);
        public List<Order> queryOrderByUserId(String userid);
        //保存订单明细
        public void saveOrderDetail(String orderNo,CartItemBean bean);

}

package com.ssmbook2020.service;
@Service("orderService")
public class OrderServiceImpl implements OrderService {
    @Autowired
    private OrderDAO orderdao = null;
    public OrderDAO getOrderdao(){
        return orderdao;
    }

    public void setOrderdao(OrderDAO orderdao){
        this.orderdao = orderdao;
    }

    /**
     * 返回订单编号
     */
```

```
    @Override
    public String addOrder(Order order){
        return orderdao.addOrder(order);
    }

    @Override
    public Order queryOrder(String orderNo){
        return orderdao.queryOrder(orderNo);
    }

    @Override
    public int updateOrder(Order order){
        //TODO Auto-generated method stub
        return 0;
    }

    @Override
    public List<Order> queryOrderByUserId(String userid){
        return orderdao.queryOrderByUserId(userid);
    }

    @Override
    public void saveOrderDetail(String orderNo,CartItemBean cartItemBean){
        orderdao.addOrderDetail(orderNo, cartItemBean);


    }

}
```

（6）定义 OrderController.java 的控制器类。这个类主要影响用户对于订单的一系列请求操作。

```
@Controller
public class OrderController {
    //省略了方法体的详细内容
    @Autowired
    private OrderService orderService = null;
    @RequestMapping(value = "/book/submitOrder.htm")
    public String handleOrder(HttpServletRequest request){
        return "submitOrderSuccess";
    }
    @RequestMapping(value = "/book/listHistoryOrder.htm")
    public String listHistoryOrder(HttpServletRequest request){
        return "listHistoryOrder";
    }

    @RequestMapping(value = "/book/listOneOrder.htm")
```

```
    public String listOneOrder(HttpServletRequest request){
        return "listOneOrder";
    }
}
```

该类的主要方法和响应的 URL 地址如表 6-4 所示。

（7）在相关的页面中组织好相关链接地址，使其能正确地请求 URL 地址，完成相应订单功能。

（8）完成订单的相关 JSP 页面。

订单模块主要有 3 个页面。第 1 个页面是 orderConfirm.jsp 页面，如图 6-9 所示。第 2 个页面是 listOneOrder.jsp 页面，作用是显示单个订单信息，如图 6-10 所示。第 3 个页面是 listHistoryOrder.jsp 页面，作用是用户查询个人历史中的所有订单信息，如图 6-11 所示。

表 6-4　OrderController 方法说明

请 求 地 址	响应方法名	返回结果页面	方 法 说 明
/book/submitOrder.htm	handleOrder()	submitOrderSuccess	处理用户提交订单，插入订单表和订单明细表
/book/listHistoryOrder.htm	listHistoryOrder()	listHistoryOrder	显示制定用户历史订单列表
/book/listOneOrder.htm	listOneOrder ()	listOneOrder	显示一个订单信息

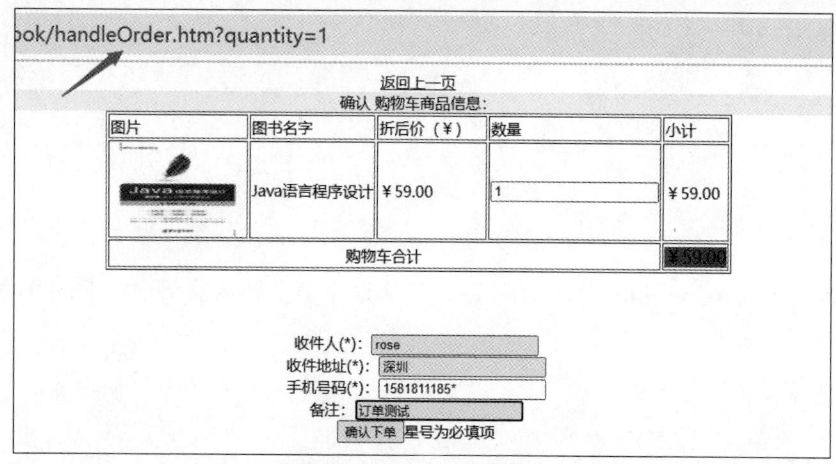

图 6-9　订单确认页面

h6/book/listOneOrder.htm?orderNo=100017

查看所有图书信息
查看我的购物车
查看我的历史订单

查询订单信息

书名：

查询

订单为:

订单编号	订单总额	订单状态	下单时间	收件人	收件地址	联系电话	运费	备注	操作
100017	¥59.00	1	2022-03-29 22:54:14	rose	深圳	19800010002	0.0	订单测试	查看详情

显示购物车

图 6-10　显示单个订单信息页面

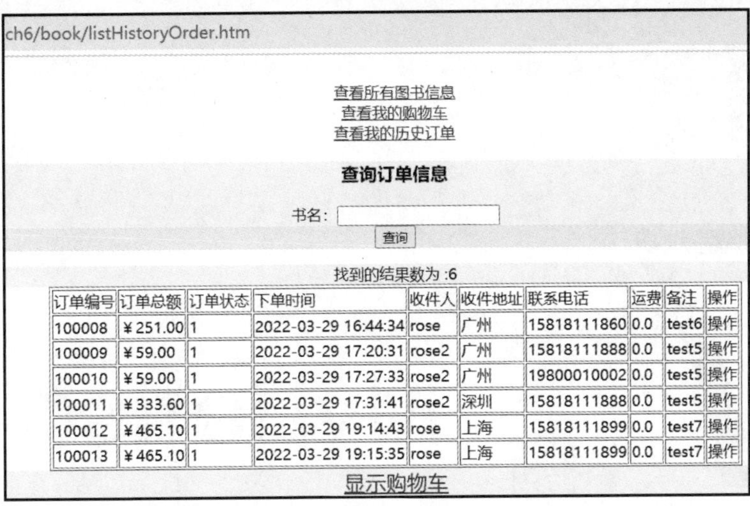

图 6-11 显示个人所有订单页面

至此，网上书店案例的主要功能基本完成。当然，还有一些细节待完善和优化，如管理员管理订单的状态和会员支付功能等。

6.5 本章小结

本章使用 Spring MVC 框架重构了网上书店案例，主要完成了图书模块、购物车模块和订单模块等功能。

习题 6

1. 请使用 Spring MVC 完成模拟发货后改变订单的状态。
2. 请完成订单状态改变后通知相应用户的功能。
3. ModelAndView 对象的常见使用方法有哪些？

第7章 MyBatis框架入门

本章学习内容

- MyBatis 框架原理；
- MyBatis 开发流程；
- MyBatis 增、删、改、查操作；
- MyBatis 结果映射；
- MyBatis 接口动态代理。

为了实现软件开发的“低耦合、高内聚”要求，通常进行分层开发，分为数据访问层、业务逻辑层、表示层三层架构，不同的层都有对应的框架，以实现该层的快速开发。例如，数据访问层有 Hibernate 和 MyBatis 等著名的框架，表示层有 Struts2 和 Spring MVC 框架。另外，Spring 框架不但可用于业务逻辑层，还可用于整合不同层的框架，最终使所有框架协调一致地工作。本章学习 MyBatis 框架的基础知识。

7.1 MyBatis 框架简介

MyBatis 是一种解决数据持久化的框架，应用于数据访问层。MyBatis 本是 Apache 的一个开源项目 iBatis，2010 年该项目由 Apache Software Foundation 迁移到 Google Code 并改名为 MyBatis，2013 年 11 月迁移到 Github。

MyBatis 是一种 ORM（Object Relational Mapping，对象关系映射）框架，用来实现 Java 对象与数据库之间的映射。Java 是面向对象的语言，而数据库是关系型的，通过 ORM 就可以实现在 Java 中用操作对象的方式操作数据库。

MyBatis 可以使用简单的 XML 或注解来配置和映射原生信息，将接口和 Java 的 POJOs（Plain Old Java Objects，普通 Java 对象）映射成数据库中的记录。

MyBatis 内部封装了 JDBC，屏蔽了 JDBC API 底层访问细节，使开发者不用与 JDBC 打交道，就可以完成对数据库的持久化操作。开发者只需要关注 SQL 语句本身，不再需要花费精力去处理加载驱动、创建连接、创建 statement 等 JDBC 原始操作所需要的繁杂过程。

MyBatis 对 JDBC 的优化和封装体现在以下几方面：

（1）使用数据库连接池对连接进行管理。

（2）SQL 语句统一存放到配置文件。

（3）SQL 语句变量和传入参数的映射及动态 SQL。

（4）动态 SQL 语句的处理。

（5）对数据库操作结果的映射和结果缓存。

（6）SQL 语句的重复。

MyBatis 的优点如下。

- 简单易学：本身体积很小且操作简单。没有任何第三方依赖，最简单的安装只要两个 JAR 文件+配置几个 SQL 映射文件。易于学习和使用，通过文档和源代码，可以比较完全地掌握它的设计思路和实现方法。

- 灵活：MyBatis 不会对应用程序或者数据库的现有设计强加任何影响。SQL 写在 XML 里，便于统一管理和优化。通过 SQL 基本上可以实现不使用数据访问框架时的所有功能，甚至更多。
- 解除 SQL 与程序代码的耦合：通过提供 DAO 层，将业务逻辑和数据访问逻辑分离，使系统的设计更清晰，更易维护，更易进行单元测试。SQL 和代码的分离，提高了系统框架的可维护性。
- 提供映射标签，支持对象与数据库的 ORM 字段关系映射。
- 提供对象关系映射标签，支持对象关系组建维护。
- 提供 XML 标签，支持编写动态 SQL。

MyBatis 的缺点如下。

- 编写 SQL 语句的工作量很大，尤其是字段多、关联表多的情况。
- SQL 语句依赖于数据库，导致数据库移植性差，不能更换数据库。
- 二级缓存机制存在问题。

视频讲解

7.2 MyBatis 开发环境

下面介绍如何下载 MyBatis 和搭建 MyBatis 开发环境。

7.2.1 MyBatis 的下载

首先到 MyBatis 中文官网 http://www.mybatis.cn/82.html 下载 mybatis-3.5.2.zip，下载后解压，解压目录如图 7-1 所示。

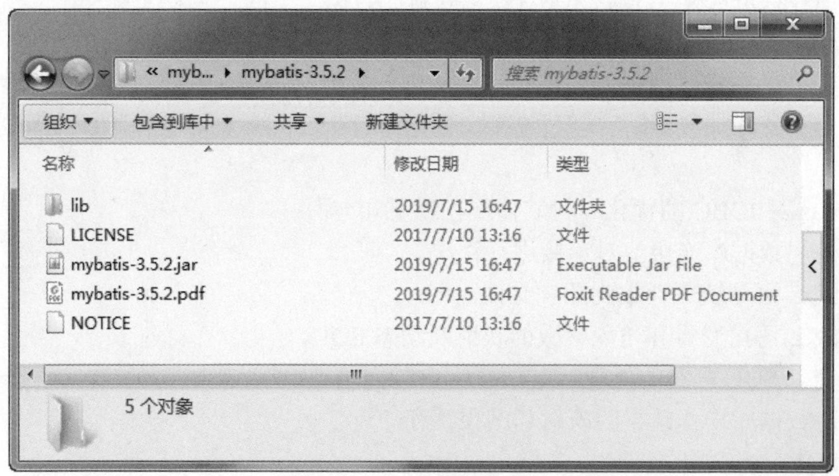

图 7-1 MyBatis 下载解压目录

其中 mybatis-3.5.2.jar 包是 MyBatis 的核心 JAR 包，mybatis-3.5.2.pdf 是 MyBatis 说明文档，lib 目录下放置的是核心 JAR 包所需的依赖 JAR 包，lib 目录打开后如图 7-2 所示。

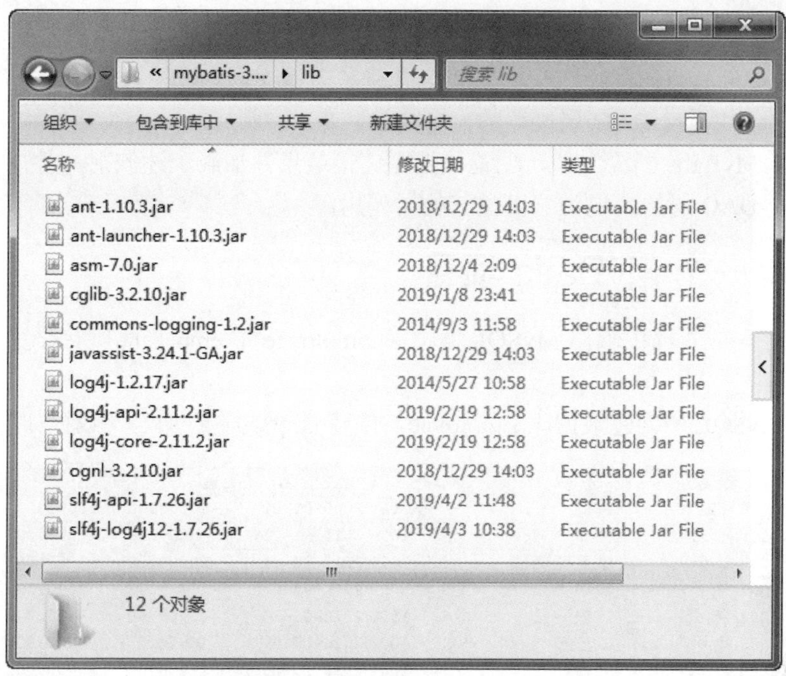

图 7-2　MyBatis 的依赖 JAR 包

7.2.2　搭建 MyBatis 开发环境

本书 MyBatis 使用 3.5.2 版，数据库使用 MySQL5.7 版，JDK 使用 1.8 版，Tomcat 使用 8.5 版，IDE 使用 Eclipse。

Java Web 项目在使用 MyBatis 前需要导入 mybatis-3.5.2.jar 包和 commons-logging-1.2.jar 包。此外，还需要连接数据库所需的 JAR 包，如连接 MySQL 数据库的 mysql-connector-java-5.1.40-bin.jar 包。在项目的 WebContent/WEB-INF/lib 目录下导入上述 JAR 包，如图 7-3 所示，选中这些 JAR 包，再右击选择 Build Path→Add to Build Path 选项即可搭建成功。

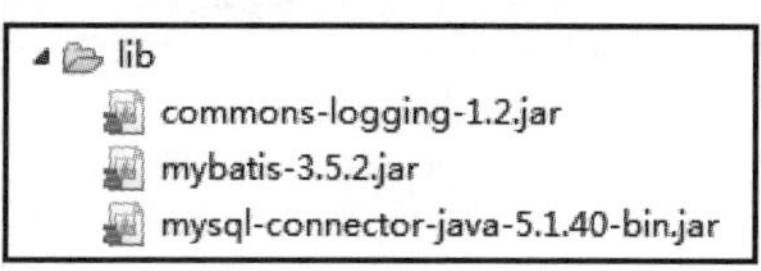

图 7-3　MyBatis 项目所需的 JAR 包

7.3　MyBatis 开发流程 

7.3.1　MyBatis 基本开发流程

搭建好开发环境后，接下来就可以进行开发了，MyBatis 基本开发流程如下。

（1）在项目中创建数据库表对应的实体类。若数据库有员工表 emp，则项目中需创建一个实体类 Emp，实体类的属性与对应表的字段相同。

（2）创建 DAO 层接口，定义增、删、改、查等操作方法。

（3）创建映射文件，其命名空间与接口全路径名称相同。在映射文件中书写 SQL 语句块，每一条语句块对应 DAO 层接口的一个方法。

（4）创建 MyBatis 配置文件，配置数据库连接信息并加载上述映射文件。

（5）创建 DAO 层的实现类，以供其他层调用。

7.3.2　第一个 MyBatis 项目

示例：在 Web 页面中输出 MySQL 数据库 Employee 中 emp 表的所有员工信息。

实现步骤：

（1）在 MySQL 中创建数据库 Employee，创建 emp 表，添加若干数据如图 7-4 所示。

empno	ename	job	mgr	hiredate	sal	comm	deptno
1000	张三丰	老板	(NULL)	2000-01-01	50000.00	10000.00	10
1001	张无忌	保安	1000	2018-01-01	6000.00	600.00	10
1002	李寻欢	项目经理	1000	2017-02-01	20000.00	5000.00	20
1003	黄飞鸿	程序员	1002	2018-02-03	10000.00	2000.00	20
1004	令狐冲	工程队长	1000	2015-01-10	18000.00	6000.00	30
1005	林冲	工程师	1004	2018-03-01	13000.00	4000.00	30
1006	鲁智深	工程师	1004	2018-10-01	11000.00	3000.00	30

图 7-4　emp 表

（2）在 Eclipse 中创建 Web 项目 mybatis1，参考本书前面章节的有关内容搭建 Spring MVC 环境，包括导入 Spring MVC 所需的 JAR 包、创建 springmvc.xml 配置文件、配置 web.xml 文件。然后在 WebContent/WEB-INF/lib 目录下添加 mybatis 所依赖的 mybatis-3.5.2.jar、commons-logging-1.2.jar、连接 MySQL 数据库所需的 mysql-connector-java-5.1.40-bin.jar、jsp 页面所需的 servlet-api.jar、jsp 页面中使用 JSTL 标签所需的 jstl-1.2.jar 等 JAR 包。项目所有 JAR 包如图 7-5 所示。

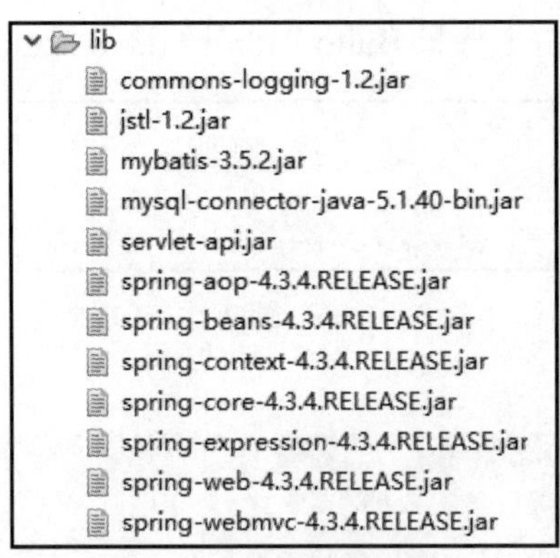

图 7-5　项目所有 JAR 包

（3）在 src 下创建 com.ssmbook2020.ch7.entity 包，包下创建与数据库表 emp 对应的实体类 Emp，关键代码如下。

```
public class Emp {
    private int empno;
```

```
    private String empname;
    private String job;
    private int manager;
    private Date hiredate;
    private double salary;
    private double commission;
    private int deptno;
    //省略构造方法和 get/set 方法
}
```

（4）新建 com.ssmbook2020.ch7.dao 包，在包下创建 IEmpDao 接口和 findAllEmps 方法，用于查找所有员工，关键代码如下。

```
public interface IEmpDao {
    public List<Emp> findAllEmps();
}
```

（5）在 com.ssmbook2020.ch7.dao 包下创建 EmpMapper.xml 映射文件，主要用于编写 SQL 语句，代码如下。

```
<?xml version="1.0" encoding="UTF-8"?>
<!DOCTYPE mapper
PUBLIC "-//mybatis.org//DTD Mapper 3.0//EN"
"http://mybatis.org/dtd/mybatis-3-mapper.dtd">
<mapper namespace="com.ssmbook2020.ch7.dao.IEmpDao">
    <select id="findAllEmps" resultType="com.ssmbook2020.ch7.entity.Emp">
        select empno,ename as empname,job,mgr as manager,hiredate,sal as
      salary,comm as commission,deptno from emp
    </select>
 </mapper>
```

映射文件有模板，可以从 Mybatis 解压目录下的 mybatis-3.5.2.pdf 文件中找到，在 PDF 文档搜索关键字 mybatis-3-mapper.dtd 即可快速找到并复制该模板。在上述代码中，<!DOCTYPE mapper PUBLIC "-//mybatis.org//DTD Mapper 3.0//EN" "http://mybatis.org/dtd/mybatis-3-mapper.dtd">是映射文件的约束信息；<mapper>标签指定映射文件的命名空间，通常命名空间与接口的全路径名称相同，本项目接口的全路径是 com.ssmbook2020.ch7.dao.IEmpDao，所以命名空间设置为<mapper namespace="com.ssmbook2020.ch7.dao.IEmpDao">；<select>标签用来设计 SQL 中的 select 查询语句，一个项目查询语句可能有多条，这里使用 id 属性来区分不同的查询语句，并且 id 属性的值与接口中的方法名相同，本例中接口中的方法名称为 findAllEmps，所以<select>标签中设置属性 id="findAllEmps"。类似地，如果一个项目还要用到 SQL 中的 insert、delete、update 语句，则映射文件中分别需要使用<insert>、<delete>、<update>标签。

<select>标签中的 resultType 属性表示将查询结果封装为该属性值指定的类型。这里将查询结果封装为 Emp 类型，注意要用全路径名称。

（6）在 src 下创建 Mybatis 的配置文件 mybatis-config.xml，代码如下。

```xml
<?xml version="1.0" encoding="UTF-8" ?>
<!DOCTYPE configuration
PUBLIC "-//mybatis.org//DTD Config 3.0//EN"
"http://mybatis.org/dtd/mybatis-3-config.dtd">
<configuration>
    <environments default="development">
        <environment id="development">
            <transactionManager type="JDBC" />
            <dataSource type="POOLED">
                <property name="driver" value="com.mysql.jdbc.Driver" />
                <property name="url"
                    value="jdbc:mysql://localhost:3306/employee" />
                <property name="username" value="root" />
                <property name="password" value="root" />
            </dataSource>
        </environment>
    </environments>
    <mappers>
        <mapper resource="com/ssmbook2020/ch7/dao/EmpMapper.xml" />
    </mappers>
</configuration>
```

该文件配置事务管理类型为 JDBC，设置了数据库连接信息，并在<mappers>标签中指出了映射文件的路径。该文件也有模板，可以从 MyBatis 解压目录下的 mybatis-3.5.2.pdf 文件中找到，在 PDF 文档搜索关键字 mybatis-3-config.dtd 即可快速找到，搜索结果如图 7-6 所示。

```xml
<?xml version="1.0" encoding="UTF-8" ?>
<!DOCTYPE configuration
  PUBLIC "-//mybatis.org//DTD Config 3.0//EN"
  "http://mybatis.org/dtd/mybatis-3-config.dtd">
<configuration>
  <environments default="development">
    <environment id="development">
      <transactionManager type="JDBC"/>
      <dataSource type="POOLED">
        <property name="driver" value="${driver}"/>
        <property name="url" value="${url}"/>
        <property name="username" value="${username}"/>
        <property name="password" value="${password}"/>
      </dataSource>
    </environment>
  </environments>
  <mappers>
    <mapper resource="org/mybatis/example/BlogMapper.xml"/>
  </mappers>
</configuration>
```

图 7-6　MyBatis 配置文件模板

将这段代码复制使用，但模板中的${}是占位符，需要替换为实际的值。${driver}代表是驱动名称，本案例用“com.mysql.jdbc.Driver”替换；${url}代表数据库 URL，替换为

"jdbc:mysql://localhost:3306/employee"；${username}指数据库用户名，替换为"root"，${password}指数据库密码，替换为"root"。

（7）在com.ssmbook2020.ch7.dao包下创建IEmpDao接口的实现类，关键代码如下。

```
public class EmpDaoImpl implements IEmpDao{
    @Override
    public List<Emp> findAllEmps() {
        SqlSession session=null;
        List<Emp> list=new ArrayList<Emp>();
        try {
            //1.读取配置文件mybatis-config.xml
            Reader reader=Resources.getResourceAsReader("mybatis-config.
xml");
            //2.创建SqlSessionFactoryBuilder对象
            SqlSessionFactoryBuilder builder=new SqlSessionFactoryBuilder();
            //3.创建SqlSessionFactory对象
            SqlSessionFactory factory=builder.build(reader);
            //4.创建SqlSession对象
            session=factory.openSession();
            //5.调用SqlSession对象的selectList方法，查找数据库,返回封装为Emp
            //对象的结果集合
    list=session.selectList("com.ssmbook2020.ch7.dao.IEmpDao.findAllEmps");
        }catch (Exception e) {
            e.printStackTrace();
        }
        return list;
    }
}
```

（8）在src下创建com.ssmbook2020.ch7.controller包,包下创建EmpController控制器，控制器下创建findAllEmps方法，调用DAO层获取数据并传递给视图。关键代码如下。

```
@Controller
public class EmpController {
    @RequestMapping("/findAll")
    public ModelAndView findAllEmps() {
        EmpDaoImpl empDaoImpl=new EmpDaoImpl();
        List<Emp> emps=empDaoImpl.findAllEmps();
        ModelAndView mv=new ModelAndView();
        mv.addObject("emps", emps);
        mv.setViewName("emplist");
        return mv;
    }
}
```

注意：本章为了让读者专注于MyBatis本身的知识，暂时未用Spring对MyBatis进行整合管理，仍然用传统的方法实现各层之间的调用（如new EmpDaoImpl()）。

（9）在 WebContent/WEB-INF 下创建 jsp 目录，jsp 目录下创建 emplist.jsp 文件，关键代码如下。

```
<body>
<table border="1">
        <tr>
            <th>编号</th><th>姓名</th><th>职位</th><th>入职日期</th>
            <th>经理</th><th>工资</th><th>奖金</th><th>部门</th>
        </tr>
        <c:forEach items="${emps }" var="emp" varStatus="vs">
            <tr ${vs.count%2==1 ? "style='background-color:yellow'" :
              "style='background-color:white'" }>
                <td>${emp.empno}</td><td>${emp.empname }</td>
                <td>${emp.job }</td><td>${emp.hiredate.toLocaleString() }
</td>
                <td>${emp.manager }</td><td>${emp.salary }</td>
                <td>${emp.commission }</td><td>${emp.deptno }</td>
            </tr>
        </c:forEach>
    </table>
</body>
```

（10）运行项目，在浏览器地址栏输入 http://localhost:8080/mybatis1/findAll，结果如图 7-7 所示。

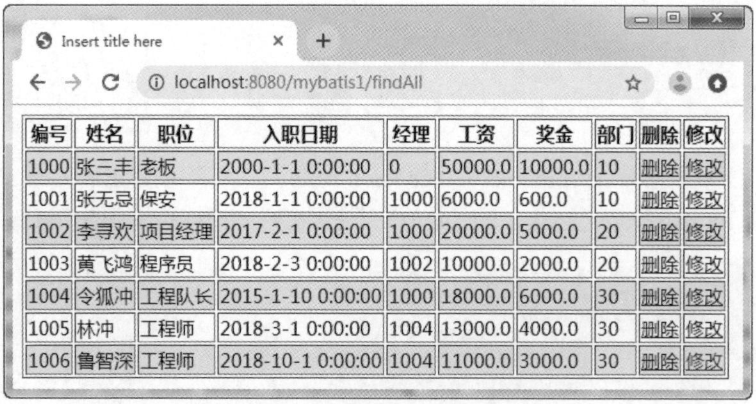

Insert title here

localhost:8080/mybatis1/findAll

编号	姓名	职位	入职日期	经理	工资	奖金	部门	删除	修改
1000	张三丰	老板	2000-1-1 0:00:00	0	50000.0	10000.0	10	删除	修改
1001	张无忌	保安	2018-1-1 0:00:00	1000	6000.0	600.0	10	删除	修改
1002	李寻欢	项目经理	2017-2-1 0:00:00	1000	20000.0	5000.0	20	删除	修改
1003	黄飞鸿	程序员	2018-2-3 0:00:00	1002	10000.0	2000.0	20	删除	修改
1004	令狐冲	工程队长	2015-1-10 0:00:00	1000	18000.0	6000.0	30	删除	修改
1005	林冲	工程师	2018-3-1 0:00:00	1004	13000.0	4000.0	30	删除	修改
1006	鲁智深	工程师	2018-10-1 0:00:00	1004	11000.0	3000.0	30	删除	修改

图 7-7　查询所有员工结果

7.3.3　MyBatis 工作流程

通过 7.3.2 节第一个 MyBatis 项目的使用，可以知道 MyBatis 的工作流程，MyBatis 的关键工作流程如下。

（1）读取 MyBatis 主配置文件，获取数据库连接信息，连接数据库。

（2）根据 MyBatis 配置文件中的映射文件路径，加载所有映射文件。

（3）根据 MyBatis 配置文件构建 SqlSessionFactory 会话工厂对象。

（4）根据上述 SqlSessionFactory 会话工厂对象创建 SqlSession 会话对象。

（5）调用 SqlSession 会话对象的操作数据库的方法。

（6）MyBatis 底层自动调用 Executor 接口操作数据库。

（7）MyBatis 底层自动映射输入参数，将 Java 对象映射到 SQL 语句中。

（8）MyBatis 底层自动将输出结果映射为 Java 对象。

7.4 使用 MyBatis 实现增、删、改、查操作

使用 MyBatis 可以实现常见的数据库增、删、改、查操作。在查询方面，可以查询单条记录，也可以查询多条记录，MyBatis 可以将单条记录封装为一个对象，将多条记录封装为对象集合。增、删、改的操作流程与查询类似，但对应映射文件中的 SQL 语句块的标签分别用<insert>、<delete>和<update>。不同的操作使用 SqlSession 的不同方法。将常用的 SqlSession 方法列表说明，如表 7-1 所示。

表 7-1 SqlSession 常用方法说明

方法名称	描　述
selectOne	查询数据库中的一条记录并封装成一个对象返回
selectList	查询数据库中的多条记录并封装成对象集合返回
insert	将对象参数传入并插入到数据库表，一个对象对应数据库表中的一条记录，返回受影响行数
delete	删除指定 id 号的记录，返回受影响行数
update	将对象参数传入并更新数据库表 ，更新的是主键与对象的 id 属性相同的记录，返回受影响行数

7.4.1 使用 selectOne 方法查询单个员工

视频讲解

7.3.2 节的案例使用了 SqlSession 的 selectList 方法，用于查询数据库中的多条记录并将查询结果封装成泛型集合。而 selectOne 方法用于查询数据库中的一条记录并将查询结果封装成单个实体对象。如果不清楚查询结果是一条还是多条，就用 selectList 方法。下面通过案例演示 selectOne 的用法。

示例：查询编号为 1001 的员工信息。

实现步骤：

（1）在 IEmpDao 接口中添加 findSingleEmp 方法，在 EmpMapper.xml 文件中添加下列 SQL 语句块。

```
<select id="findSingleEmp" resultType="com.ssmbook2020.ch7.entity.Emp">
select empno,ename as empname,job,mgr as manager,hiredate,sal as
salary,comm as commission,deptno from emp where empno=1001
</select>
```

（2）在 EmpDaoImpl 接口实现类中创建 findSingleEmp 方法，关键代码如下。

```
@Override
public Emp findSingleEmp() {
     SqlSession session=null;
     Emp emp=null;
     try {
          //1.读取配置文件 mybatis-config.xml
          Reader reader=Resources.getResourceAsReader("mybatis-config.xml");
          //2.创建 SqlSessionFactoryBuilder 对象
          SqlSessionFactoryBuilder builder=new SqlSessionFactoryBuilder();
          //3.创建 SqlSessionFactory 对象
          SqlSessionFactory factory=builder.build(reader);
          //4.创建 SqlSession 对象
          session=factory.openSession();
          //5.调用 SqlSession 对象的 selectOne 方法，查找数据库的一条记录,将结果映射
          //为 Emp 对象
          emp=session.selectOne("com.ssmbook2020.ch7.dao.IEmpDao.
findSingleEmp");
     }catch (Exception e) {
            e.printStackTrace();
     }
     return emp;
}
```

（3）在 EmpController 中添加方法，关键代码如下。

```
@RequestMapping("/findSingle")
public ModelAndView findSingleEmp() {
     EmpDaoImpl empDaoImpl=new EmpDaoImpl();
     Emp emp=empDaoImpl.findSingleEmp();
     ModelAndView mv=new ModelAndView();
     mv.addObject("emp", emp);
     mv.setViewName("emp");
     return mv;
}
```

（4）在 WebContent/WEB-INF/jsp 目录下创建 emp.jsp 页面，关键代码如下。

```
<body>
<h1>员工信息</h1>
<table border="1">
    <tr><th>编号</th><td>${emp.empno }</td></tr>
    <tr><th>姓名</th><td>${emp.empname}</td>    </tr>
    <tr><th>职位</th><td>${emp.job }</td>    </tr>
    <tr><th>入职日期</th><td>${emp.hiredate.toLocaleString()}</td> </tr>
    <tr><th>经理</th><td>${emp.manager }</td>    </tr>
    <tr><th>工资</th><td>${emp.salary }</td>    </tr>
```

```
    <tr><th>奖金</th><td>${emp.commission }</td>    </tr>
    <tr><th>部门</th><td>${emp.deptno }</td>    </tr>
</table>
</body>
```

（5）运行测试，在浏览器地址栏中输入 http://localhost:8080/mybatis1/findSingle，结果如图 7-8 所示。

图 7-8　查询单个员工结果

7.4.2　使用 insert 方法添加员工

视频讲解

SqlSession 中的 insert 方法用于添加记录到数据库，返回受影响的行数（int 类型）。该方法需要用到参数，通常参数设置为实体对象。MyBatis 默认事务管理方式为“关闭自动提交”，所以需要手动提交 insert、delete、update 这 3 个数据库 SQL 操作。

示例：在项目 mybatis1 中实现添加新员工。

实现步骤：

（1）在 IEmpDao 接口中添加 insertEmp(Emp emp) 方法，然后在 EmpMapper.xml 映射文件中添加 insert 语句块如下。

```
<insert id="insertEmp" parameterType="com.ssmbook2020.ch7.entity.Emp">
    insert into emp(empno,ename,job,mgr,hiredate,sal,comm,deptno)
    values(#{empno},#{empname},#{job},#{manager},#{hiredate},#{salary},
    #{commission},#{deptno})
</insert>
```

其中，parameterType 用于指定输入参数类型，也可以不指定，系统会自动识别，这里表示传入的参数类型是 Emp 类型；#{}代表一个参数，是一个占位符，里面的参数名称与传入的对象的属性名称一致。

（2）在接口的实现类中添加 insertEmp(Emp emp)方法，代码如下。

```
@Override
```

```
public int insertEmp(Emp emp) {
    SqlSession session=null;
    int count=0;
    try {
        //1.读取配置文件 mybatis-config.xml
        Reader reader=Resources.getResourceAsReader("mybatis-config.xml");
        //2.创建 SqlSessionFactoryBuilder 对象
        SqlSessionFactoryBuilder builder=new SqlSessionFactoryBuilder();
        //3.创建 SqlSessionFactory 对象
        SqlSessionFactory factory=builder.build(reader);
        //4.创建 SqlSession 对象
        session=factory.openSession();
        //5.调用 SqlSession 对象的 insert 方法，插入数据库，返回受影响行数
    count=session.insert("com.ssmbook2020.ch7.dao.IEmpDao.insertEmp",emp);
        session.commit();//6.手动提交事务
    }catch (Exception e) {
        e.printStackTrace();
    }
    return count;
}
```

（3）在 EmpController 控制器中添加 insertEmp 方法，代码如下。

```
@RequestMapping("/insertEmp")
public ModelAndView insertEmp(Emp emp) {
    EmpDaoImpl empDaoImpl=new EmpDaoImpl();
    empDaoImpl.insertEmp(emp);
    ModelAndView mv=new ModelAndView();
    mv.setViewName("forward:findAll");//转发到 findAll，重新查找员工
    return mv;
}
```

（4）在 WebContent 下创建 add.jsp 页面，关键代码如下。

```
<body>
<h1>添加员工</h1>
<form action="insertEmp" method="post">
<table>
   <tr><th>编号</th><td><input type="text" name="empno"/></td></tr>
   <tr><th>姓名</th><td><input type="text" name="empname"/></td> </tr>
   <tr><th>职位</th><td><input type="text" name="job"/></td> </tr>
   <tr><th>入职日期</th><td><input type="text" name="hiredate"
                            placeholder="格式:yyyy/mm/dd"/></td> </tr>
   <tr><th>经理</th><td><input type="text" name="manager"/></td> </tr>
   <tr><th>工资</th><td><input type="text" name="salary"/></td> </tr>
   <tr><th>奖金</th><td><input type="text" name="commission"/></td> </tr>
   <tr><th>部门</th><td><input type="text" name="deptno"/></td> </tr>
   <tr><td colspan=2 align="center"><input type="submit" value="确定"/>
```

```
</td></tr>
    </table>
    </form>
    </body>
```

（5）在 emplist.jsp 上添加页面 add.jsp 的链接，运行测试，打开添加页面并输入数据，如图 7-9 所示（注意日期用 yyyy/mm/dd 格式）。

图 7-9　添加员工界面

单击确定后，结果如图 7-10 所示。由此可见员工信息中多了一条记录。

Insert title here

localhost:8080/mybatis1/insertEmp

添加员工

编号	姓名	职位	入职日期	经理	工资	奖金	部门
1000	张三丰	老板	2000-1-1 0:00:00	0	50000.0	10000.0	10
1001	张无忌	保安	2018-1-1 0:00:00	1000	6000.0	600.0	10
1002	李寻欢	项目经理	2017-2-1 0:00:00	1000	20000.0	5000.0	20
1003	黄飞鸿	程序员	2018-2-3 0:00:00	1002	10000.0	2000.0	20
1004	令狐冲	工程队长	2015-1-10 0:00:00	1000	18000.0	6000.0	30
1005	林冲	工程师	2018-3-1 0:00:00	1004	13000.0	4000.0	30
1006	鲁智深	工程师	2018-1-1 0:00:00	1004	11000.0	3000.0	30
1007	李白	教授	2018-1-1 0:00:00	1000	10000.0	3000.0	30

图 7-10　添加员工结果

7.4.3　使用 delete 方法删除员工

视频讲解

SqlSession 的 delete 方法实现删除功能，返回受影响的行数（int 类型）。

示例：在 mybatis1 项目中实现删除一条记录的功能。

实现步骤：

（1）在 IEmpDao 接口中添加 deleteEmp(int empno) 方法，然后在对应的 EmpMapper.xml 映射文件中添加 delete 语句块如下。

```
<delete id="deleteEmp" parameterType="int">
    delete from emp where
    empno=#{empno}
</delete>
```

parameterType="int"表示传入的参数类型是一个整型。

（2）在 EmpDao 接口实现类中添加 deleteEmp 方法，代码如下。

```
@Override
public int deleteEmp(int empno) {
    SqlSession session=null;
    int count=0;
    try {
        //1.读取配置文件 mybatis-config.xml
        Reader reader=Resources.getResourceAsReader("mybatis-config.xml");
        //2.创建 SqlSessionFactoryBuilder 对象
        SqlSessionFactoryBuilder builder=new SqlSessionFactoryBuilder();
        //3.创建 SqlSessionFactory 对象
        SqlSessionFactory factory=builder.build(reader);
        //4.创建 SqlSession 对象
        session=factory.openSession();
        //5.调用 SqlSession 对象的 delete 方法，删除指定编号的员工，返回受影响的行数
        count=session.delete("com.ssmbook2020.ch7.dao.IEmpDao.
                            deleteEmp",empno);
        session.commit();
    }catch (Exception e) {
        e.printStackTrace();
    }
    return count;
}
```

（3）在 EmpController 控制器中添加以下代码。

```
@RequestMapping("/deleteEmp")
public ModelAndView deleteEmp(int empno) {
    EmpDaoImpl empDaoImpl=new EmpDaoImpl();
    empDaoImpl.deleteEmp(empno);
    ModelAndView mv=new ModelAndView();
    mv.setViewName("forward:findAll");//转发到 findAll，重新查找员工
    return mv;
}
```

（4）修改 emplist.jsp，添加一个列用于删除，关键代码如下。

```
    <td><a href="deleteEmp?empno=${emp.empno}" onclick="return confirm('你确定要删除吗?')">删除</a></td>
```

（5）测试运行。在浏览器地址栏中输入 http://localhost:8080/mybatis1/findAll，然后单击列表中编号为 1007 的员工的“删除”超链接，如图 7-11 所示。

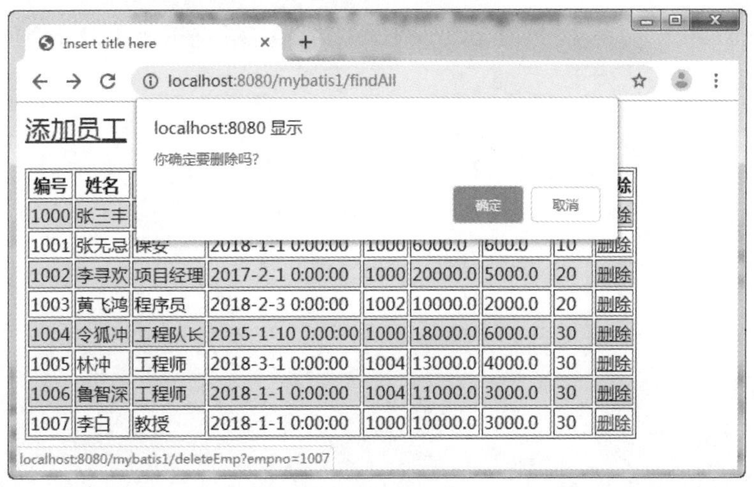

图 7-11　删除操作

在图 7-11 所示的对话框中单击确定，结果如图 7-12 所示，显示编号为 1007 的员工删除成功了。

Insert title here

localhost:8080/mybatis1/deleteEmp?empno=1007

添加员工

编号	姓名	职位	入职日期	经理	工资	奖金	部门	删除	修改
1000	张三丰	老板	2000-1-1 0:00:00	0	50000.0	10000.0	10	删除	修改
1001	张无忌	保安	2018-1-1 0:00:00	1000	6000.0	600.0	10	删除	修改
1002	李寻欢	项目经理	2017-2-1 0:00:00	1000	20000.0	5000.0	20	删除	修改
1003	黄飞鸿	程序员	2018-2-3 0:00:00	1002	10000.0	2000.0	20	删除	修改
1004	令狐冲	工程队长	2015-1-10 0:00:00	1000	18000.0	6000.0	30	删除	修改
1005	林冲	工程师	2018-3-1 0:00:00	1004	13000.0	4000.0	30	删除	修改
1006	鲁智深	工程师	2018-10-1 0:00:00	1004	11000.0	3000.0	30	删除	修改

图 7-12　删除结果

7.4.4　使用 update 方法修改员工

视频讲解

SqlSession 的 update 方法用于修改数据库记录，返回受影响行数。

示例：修改一个员工的信息，修改前查询并显示原来的信息。

（1）在 IEmpDao 接口中添加 findEmpById(int empno)方法，然后在对应的 EmpMapper.xml 映射文件中添加 select 语句块如下。

```
    <select id="findEmpById" parameterType="int"
```

```
    resultType="com.ssmbook2020.ch7.entity.Emp">
        select empno,ename as empname,job,mgr as
        manager,hiredate,sal as salary,comm as commission,deptno from emp
        where empno=#{empno}
    </select>
```

再在 IEmpDao 中添加 updateEmp(Emp emp) 方法，然后在对应的 EmpMapper.xml 映射文件中添加 select 语句块如下。

```
    <update id="updateEmp" parameterType="com.ssmbook2020.ch7.entity.Emp">
          update emp set
          ename=#{empname},job=#{job},mgr=#{manager},hiredate=#{hiredate},
          sal=#{salary},comm=#{commission},deptno=#{deptno} where empno=#{empno}
    </update>
```

（2）在 EmpDao 接口实现类中添加 findEmpById 方法，代码如下。

```
    @Override
    public Emp findEmpById(int empno) {
        SqlSession session=null;
        Emp emp=null;
        try {
            //1.读取配置文件 mybatis-config.xml
            Reader reader=Resources.getResourceAsReader("mybatis-config.xml");
            //2.创建 SqlSessionFactoryBuilder 对象
            SqlSessionFactoryBuilder builder=new SqlSessionFactoryBuilder();
            //3.创建 SqlSessionFactory 对象
            SqlSessionFactory factory=builder.build(reader);
            //4.创建 SqlSession 对象
            session=factory.openSession();
            //5.调用 SqlSession 对象的 selectOne 方法，查找数据库一条记录，将结果映射
            //为 Emp 对象
            emp=session.selectOne("com.ssmbook2020.ch7.dao.IEmpDao.
findEmpById",empno);
        }catch (Exception e) {
             e.printStackTrace();
        }
        return emp;
    }
```

再添加 updateEmp 方法，代码如下。

```
    @Override
    public int updateEmp(Emp emp) {
        SqlSession session=null;
        int count=0;
        try {
            //1.读取配置文件 mybatis-config.xml
```

```
        Reader reader=Resources.getResourceAsReader("mybatis-config.xml");
        //2.创建 SqlSessionFactoryBuilder 对象
        SqlSessionFactoryBuilder builder=new SqlSessionFactoryBuilder();
        //3.创建 SqlSessionFactory 对象
        SqlSessionFactory factory=builder.build(reader);
        //4.创建 SqlSession 对象
        session=factory.openSession();
        //5.调用 SqlSession 对象的 update 方法，修改数据库,返回受影响行数
    count=session.update("com.ssmbook2020.ch7.dao.IEmpDao.updateEmp",emp);
        session.commit();
    }catch (Exception e) {
        e.printStackTrace();
    }
    return count;
}
```

（3）在 EmpController 控制器中添加以下代码。

```
@RequestMapping("/findEmpById")
public ModelAndView findEmpById(int empno) {
    EmpDaoImpl empDaoImpl=new EmpDaoImpl();
    Emp emp=empDaoImpl.findEmpById(empno);
    ModelAndView mv=new ModelAndView();
    mv.addObject("emp", emp);
    mv.setViewName("update");
    return mv;
}

@RequestMapping("/updateEmp")
public ModelAndView updateEmp(Emp emp) {
    EmpDaoImpl empDaoImpl=new EmpDaoImpl();
    empDaoImpl.updateEmp(emp);
    ModelAndView mv=new ModelAndView();
    mv.setViewName("forward:findAll");//转发到 findAll，重新查找员工
    return mv;
}
```

（4）修改 emplist.jsp，添加一个列用于修改，关键代码如下。

```
<td><a href="findEmpById?empno=${emp.empno}">修改</a></td>
```

（5）在 WebContent/WEB-INF/jsp 目录下创建 update.jsp，关键代码如下。

```
<body>
<h1>修改员工</h1>
<form action="updateEmp" method="post">
<table>
    <tr><th>编号</th><td><input type="text" name="empno" value="${emp.empno}"
        readonly="readonly"/></td></tr>
```

```
        <tr><th>姓名</th><td><input type="text" name="empname"
            value="${emp.empname}"/></td></tr>
        <tr><th>职位</th><td><input type="text" name="job" value="${emp.job}"/>
         </td></tr>
        <tr><th>入职日期</th><td><input type="text" name="hiredate"
            value="${emp.hiredate.toLocaleString}"/>（必须改为 yyyy/mm/dd 格式）
        </td></tr>
        <tr><th>经理</th><td><input type="text" name="manager"
            value="${emp.manager}"/></td></tr>
        <tr><th>工资</th><td><input type="text" name="salary"
            value="${emp.salary}"/></td></tr>
        <tr><th>奖金</th><td><input type="text" name="commission"
            value="${emp.commission}"/></td></tr>
        <tr><th>部门</th><td><input type="text" name="deptno"
            value="${emp.deptno}"/></td></tr>
        <tr><td colspan=2 align="center"><input type="submit" value="确定"/>
        </td></tr>
    </table>
    </form>
    </body>
```

（6）运行测试，在浏览器地址栏中输入 http://localhost:8080/mybatis1/findAll，然后单击列表中编号为 1006 的员工的“修改”超链接，弹出如图 7-13 所示的修改界面。

图 7-13　修改员工前

修改姓名为“武松”，职位为“经理”，入职日期改为“2018/10/1”（必须这种日期格式），工资改为“20000”，如图 7-14 所示。

单击“确定”按钮，结果如图 7-15 所示。

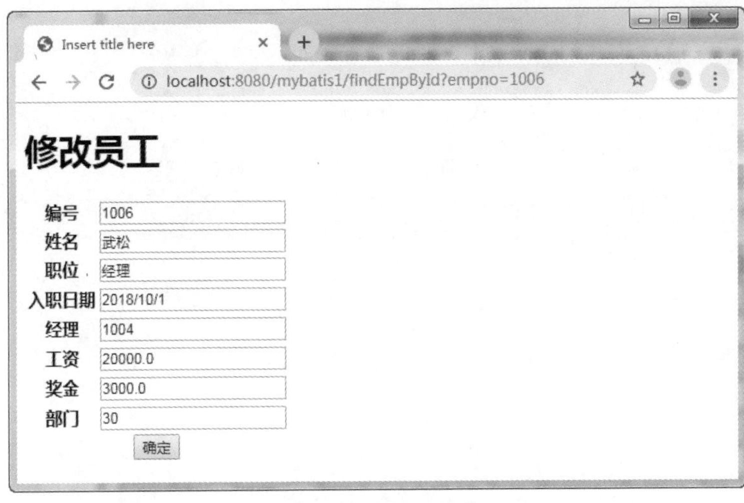

图 7-14　修改为新值

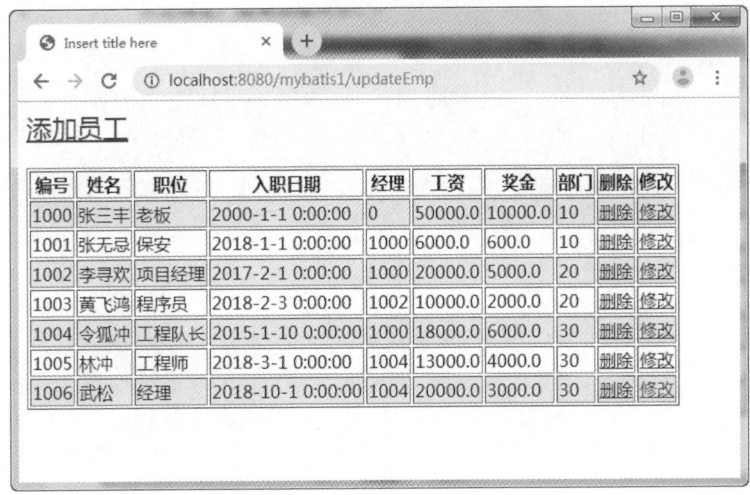

图 7-15　修改结果

由此可见，员工信息修改成功了。

7.4.5　使用工具类 MyBatisUtil 减少冗余

视频讲解

在上面的案例中，DAO 层实现类的每个方法的前面内容是相同的，都是用来获取 SqlSession 对象，代码冗余度大。可以将这些代码封装为一个类，再调用有关方法，即可大大减少冗余。MyBatisUtil 正是这样的一个类，里面有获取 SqlSession 对象的方法和关闭 SqlSession 对象的方法，该类的关键代码如下。

```
public class MyBatisUtil {
    private MyBatisUtil(){
    }
    private static final String RESOURCE = "mybatis-config.xml";
    private static SqlSessionFactory sqlSessionFactory = null;
    private static ThreadLocal<SqlSession> threadLocal = new
```

```
ThreadLocal<SqlSession>();
    static {
        Reader reader = null;
        try {
            reader = Resources.getResourceAsReader(RESOURCE);
            SqlSessionFactoryBuilder builder = new SqlSessionFactoryBuilder();
            sqlSessionFactory = builder.build(reader);
        } catch (Exception e1) {
            e1.printStackTrace();
            throw new ExceptionInInitializerError("初始化 MyBatis 错误");
        }
    }
    public static SqlSessionFactory getSqlSessionFactory(){
        return sqlSessionFactory;
    }
    public static SqlSession getSession(){
        SqlSession session = threadLocal.get();
        if (session == null){
            session =
            (sqlSessionFactory !=null) ?sqlSessionFactory.openSession():null;
            threadLocal.set(session);
        }
        return session;
    }
    public static void closeSession(){
        SqlSession session = (SqlSession) threadLocal.get();
        threadLocal.set(null);
        if (session !=null){
            session.close();
        }
    }
}
```

将该类创建到 com.ssmbook2020.ch7.utils 包下，接下来对 EmpDaoImpl 类进行简化，简化后的关键代码如下。

```
public class EmpDaoImpl implements IEmpDao{
    @Override
    public List<Emp> findAllEmps() {
        SqlSession session=null;
        List<Emp> list=new ArrayList<Emp>();
        try {
            session=MyBatisUtil.getSession();
            list=session.selectList("com.ssmbook2020.ch7.dao.IEmpDao.
findAllEmps");
        }catch (Exception e) {
            e.printStackTrace();
```

```
        }
        return list;
    }

    @Override
    public Emp findSingleEmp() {
        SqlSession session=null;
        Emp emp=null;
        try {
            session=MyBatisUtil.getSession();
            emp=session.selectOne("com.ssmbook2020.ch7.dao.IEmpDao.
findSingleEmp");
        }catch (Exception e) {
            e.printStackTrace();
        }
        return emp;
    }

    @Override
    public int insertEmp(Emp emp) {
        SqlSession session=null;
        int count=0;
        try {
            session=MyBatisUtil.getSession();
            count=session.insert("com.ssmbook2020.ch7.dao.IEmpDao.
insertEmp",emp);
            session.commit();
        }catch (Exception e) {
            e.printStackTrace();
        }
        return count;
    }

    @Override
    public int deleteEmp(int empno) {
        SqlSession session=null;
        int count=0;
        try {
            session=MyBatisUtil.getSession();
            count=session.delete("com.ssmbook2020.ch7.dao.IEmpDao.
deleteEmp",empno);
            session.commit();
        }catch (Exception e) {
            e.printStackTrace();
        }
        return count;
    }
```

```
    @Override
    public Emp findEmpById(int empno) {
        SqlSession session=null;
        Emp emp=null;
        try {
            session=MyBatisUtil.getSession();
            emp=session.selectOne("com.ssmbook2020.ch7.dao.IEmpDao.
findEmpById",empno);
        }catch (Exception e) {
            e.printStackTrace();
        }
        return emp;
    }

    @Override
    public int updateEmp(Emp emp) {
        SqlSession session=null;
        int count=0;
        try {
            session=MyBatisUtil.getSession();
            count=session.update("com.ssmbook2020.ch7.dao.IEmpDao.
updateEmp",emp);
            session.commit();
        }catch (Exception e) {
            e.printStackTrace();
        }
        return count;
    }
}
```

由此可见，代码大大简化了，后面的案例都使用该工具类进行简化。

视频讲解

7.5 parameterType 输入参数

在 7.4 节的案例中多次用到了输入参数，输入参数可以是基本类型、对象类型，也可以是 Map 类型，还可以是数组或集合类型。

若是基本类型，则只能传递一个参数。SQL 语句块中的 parameterType 属性设置为基本类型，如 parameterType="int"，也可以不设置 parameterType 属性，系统会自动判断。SQL 语句中用#{}占位符表示该参数，#{}里面可以用与参数名称相同或不同的字符串。例如，接口方法为 deleteEmp(int empno)，SQL 语句为 delete from emp where empno=#{empno}，这里用到了#{empno}，#{ }里面的字符串正好与参数相同，但改为不同的#{id}也是可以的。输入参数为基本类型的具体用法参见 7.4.2 节的案例，其输入参数是员工编号 empno，参数类型是整型。

若是对象类型，则可以将多个参数封装到一个对象中，以对象整体作为一个输入参数。将parameterType属性设置为该对象，如parameterType="com.ssmbook2020.ch7.entity.Emp"，SQL语句中的各个占位符#{}里面的字符串必须与对象的各个属性名称相同。参数类型为对象类型的参见7.4.1节的案例，输入参数为Emp类型。

如果多个参数可以封装为一个对象，就用对象作为参数，但如果不能封装为对象，则可以使用Map类型作为参数。Map类型作为参数的话，paramaterType可以设置为parameterType="java.util.Map"，SQL语句中的#{}占位符里面的字符串必须与Map集合中的key相同。下面举例说明如何设置Map作为参数，另外参数为集合或数组类型的参见8.1.4节的案例。

示例： 查询某个部门工资在xxx与xxx之间的员工信息，如查询30号部门的工资在10000~15000的员工。

分析： 该查询操作需要用到部门编号、最低工资和最高工资3个参数，而基本类型只能用一个参数，显然无法满足要求。另外，这3个参数也无法封装为一个Emp对象（Emp对象没有“最高工资”与“最低工资”这些属性，只有“工资”属性）。最好的办法是将这3个参数封装到Map中。

实现步骤：

（1）在项目MyBatis1中的IEmpDao接口中添加searchEmpsByMap(Map map)方法，在EmpDao.xml映射文件中添加如下代码块。

```
<select id="searchEmpsByMap" parameterType="java.util.Map"
resultType="com.ssmbook2020.ch7.entity.Emp">
    select empno,ename as empname,job,mgr as
    manager,hiredate,sal as salary,comm as commission,deptno from emp
    where deptno=#{deptno} and sal between #{minsal} and #{maxsal}
</select>
```

这里指定了输入参数类型为java.util.Map，Mybatis将自动读取输入参数Map中的各个键值到SQL语句中与键名相同的占位符中。

（2）在EmpDapImpl接口实现类中添加如下代码。

```
@Override
public List<Emp> searchEmpsByMap(Map map) {
    SqlSession session=null;
    List<Emp> list=new ArrayList<Emp>();
    try {
        session=MyBatisUtil.getSession();
        list=session.selectList("com.ssmbook2020.ch7.dao.IEmpDao.
searchEmpsByMap",map);
    }catch (Exception e) {
        e.printStackTrace();
    }
    return list;
```

```
}
```

（3）在 EmpController 控制器中添加如下代码。

```
@RequestMapping("/searchEmpsByMap")
public ModelAndView searchEmpsByMap(int deptno,double minsal,double maxsal) {
    EmpDaoImpl empDaoImpl=new EmpDaoImpl();
    Map map=new HashMap();
    map.put("deptno", deptno);
    map.put("minsal", minsal);
    map.put("maxsal", maxsal);

    List<Emp> emps=empDaoImpl.searchEmpsByMap(map);
    ModelAndView mv=new ModelAndView();
    mv.addObject("emps", emps);
    mv.setViewName("emplist");
    return mv;
}
```

（4）在 WebContent 下创建 search.jsp，关键代码如下。

```
<body>
<form action="searchEmpsByMap" method="post">
    部门编号：<input type="text" name="deptno" /> <br/>
    最低工资：<input type="text" name="minsal" /> <br/>
    最高工资：<input type="text" name="maxsal" /> <br/>
<input type="submit" value="查询员工" /> <br/>
</form>
</body>
```

（5）运行测试，在浏览器地址栏中输入 http://localhost:8080/mybatis1/search.jsp，在查询页面中输入查询数据，如图 7-16 所示。

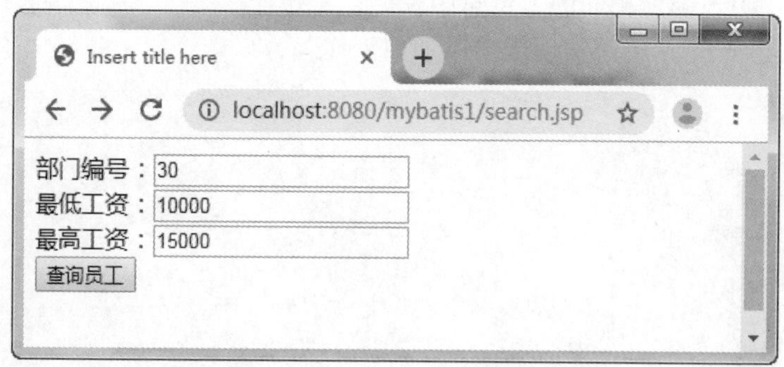

图 7-16　工资查询界面

工资查询结果如图 7-17 所示。

编号	姓名	职位	入职日期	经理	工资	奖金	部门	删除	修改
1005	林冲	工程师	2018-3-1 0:00:00	1004	13000.0	4000.0	30	删除	修改
1006	鲁智深	工程师	2018-1-1 0:00:00	1004	11000.0	3000.0	30	删除	修改
1007	李白	教授	2018-1-1 0:00:00	1000	10000.0	3000.0	30	删除	修改

图 7-17　工资查询结果

7.6 ResultMap 结果映射

视频讲解

在 7.5 节的案例中，SQL 的 select 查询语句中存在别名，如 ename as empname、sal as salary，这是因为数据库表中的字段名与实体类的属性名不同。从数据库中查询出来的结果要封装为指定的实体类，就需要使用别名使数据库表中的字段名与实体类的属性名“一致”。如果不使用别名，情况会怎样呢？下面对此进行测试。

（1）在 IEmpDao 接口中新建 findAllEmpsResultMap 方法，对应地在 EmpMapper.xml 映射文件中添加 select 语句块如下。

```
<select id="findAllEmpsResultMap" resultType="com.ssmbook2020.ch7.entity
.Emp">
        select * from emp
</select>
```

显然这里的 SQL 没用到别名。

（2）在 EmpDaoImpl 接口实现类中创建 findAllEmpsResultMap 方法，代码如下。

```
@Override
public List<Emp> findAllEmpsResultMap() {
    SqlSession session=null;
    List<Emp> list=new ArrayList<Emp>();
    try {
        session=MyBatisUtil.getSession();
         list=session.selectList("com.ssmbook2020.ch7.dao.IEmpDao.
findAllEmpsResultMap");
    }catch (Exception e) {
        e.printStackTrace();
    }
    return list;
}
```

（3）在 EmpController 控制器中添加如下代码。

```
@RequestMapping("/findAllEmpsResultMap")
public ModelAndView findAllEmpsResultMap() {
    EmpDaoImpl empDaoImpl=new EmpDaoImpl();
    List<Emp> emps=empDaoImpl.findAllEmpsResultMap();
    ModelAndView mv=new ModelAndView();
    mv.addObject("emps", emps);
    mv.setViewName("emplist");
    return mv;
}
```

（4）运行测试，在浏览地址栏中输入 http://localhost:8080/mybatis1/findAllEmpsResultMap，结果如图 7-18 所示，可以发现姓名等多个列没有数据了。

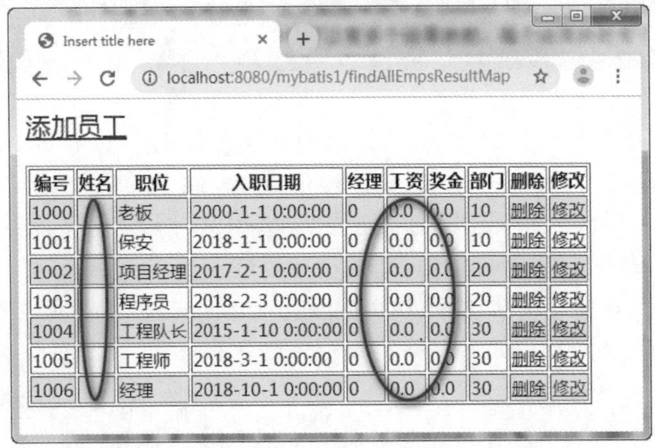

图 7-18　缺失部分值

显然数据库表 emp 的字段与实体类 Emp 属性不一致的地方都没数据。这是因为 select 语句查询出来的有些列名与实体类的属性名不一致，导致无法将这些列的数据封装到指定对象的属性中。这个问题除了可以用 SQL 语句的别名解决外，也可以使用 ResultMap 结果映射来解决。使用 ResultMap 结果映射解决问题的步骤如下。

（1）在 EmpMapper.xml 映射文件中添加<resultMap>节点，设置结果映射，代码如下所示。

```
<!-- 配置结果映射 -->
<resultMap id="empResultMap" type="com.ssmbook2020.ch7.entity.Emp">
    <id property="empno" column="empno" />
    <result property="empname" column="ename" />
    <result property="job" column="job" />
    <result property="manager" column="mgr" />
    <result property="hiredate" column="hiredate" />
    <result property="salary" column="sal" />
    <result property="commission" column="comm" />
    <result property="deptno" column="deptno" />
</resultMap>
```

这个配置就是结果映射，它将数据库表中的 column（列）与实体类中的 property（属性）一个一个地进行对应（映射），包括与属性名不一致的列名。这样数据库表中的所有 column（列）都能在封装目标对象中找到对应的 property（属性）进行对应。一个映射文件中可以配置多个结果映射，每个结果映射都有 id 号来标识。<select>语句块若需要使用某个结果映射，就引用对应的 id。<select>语句块一旦使用了某个结果映射，则其 SQL 语句查询出来的结果就用这个结果映射进行封装。具体引用方法见步骤（2）。

（2）将 EmpMapper.xml 映射文件中的原<select>语句块修改为如下代码。

```
<select id="findAllEmpsResultMap"  resultMap="empResultMap">
     select * from emp
</select>
```

其中最主要的修改是将原来的 resultType="com.ssmbook2020.ch7.entity.Emp" 改为 resultMap="empResultMap"。表示查询结果不是直接封装为 Emp 类型，而是使用指定的某个结果映射进行封装，其中的 empResultMap 对应步骤（1）设置的结果映射的 id 号。

（3）重新运行测试，结果如图 7-19 所示，显然数据都正确显示了。

Insert title here

localhost:8080/mybatis1/findAllEmpsResultMap

添加员工

编号	姓名	职位	入职日期	经理	工资	奖金	部门	删除	修改
1000	张三丰	老板	2000-1-1 0:00:00	0	50000.0	10000.0	10	删除	修改
1001	张无忌	保安	2018-1-1 0:00:00	1000	6000.0	600.0	10	删除	修改
1002	李寻欢	项目经理	2017-2-1 0:00:00	1000	20000.0	5000.0	20	删除	修改
1003	黄飞鸿	程序员	2018-2-3 0:00:00	1002	10000.0	2000.0	20	删除	修改
1004	令狐冲	工程队长	2015-1-10 0:00:00	1000	18000.0	6000.0	30	删除	修改
1005	林冲	工程师	2018-3-1 0:00:00	1004	13000.0	4000.0	30	删除	修改
1006	武松	经理	2018-10-1 0:00:00	1004	20000.0	3000.0	30	删除	修改

图 7-19　数据完整显示

ResultMap 可以灵活地解决 SQL 查询结果与封装对象之间的映射，并且能深入地应用到多表查询中。

7.7 接口动态代理

视频讲解

在前面的案例中，每个 DAO 层的接口都有对应的实现类，但仔细观察可以发现这些实现类似乎有些“多余”。因为方法名称在接口定义，SQL 语句则在 Mapper 映射文件中定义，感觉实现类中的方法没有做什么“实质性”的工作。不要 DAO 层的实现类而实现同样的功能是否可行？

可以不用手动实现 DAO 层的接口，MyBatis 框架将根据接口定义创建接口的动态代理对象，代理对象的方法体实现 Mapper 接口中定义的方法。只需要在使用的时候通过 SqlSession 对象的 getMapper 方法即可获取对应 DAO 接口的代理对象，再调用该方法即可。

示例：查询指定编号的员工信息。

实现步骤：

（1）新建一个 IEmpDao2 接口，定义一个抽象方法，代码如下。

```
public interface IEmpDao2 {
    public Emp findEmpByIdGetMapper(int empno);
}
```

（2）新建 EmpMapper2.xml 映射文件，代码如下。

```
<?xml version="1.0" encoding="UTF-8"?>
<!DOCTYPE mapper
PUBLIC "-//mybatis.org//DTD Mapper 3.0//EN"
"http://mybatis.org/dtd/mybatis-3-mapper.dtd">
<mapper namespace="com.ssmbook2020.ch7.dao.IEmpDao2">
    <!-- 配置结果映射 -->
    <resultMap id="empResultMap" type="com.ssmbook2020.ch7.entity.Emp">
        <id property="empno" column="empno" />
        <result property="empname" column="ename" />
        <result property="job" column="job" />
        <result property="manager" column="mgr" />
        <result property="hiredate" column="hiredate" />
        <result property="salary" column="sal" />
        <result property="commission" column="comm" />
        <result property="deptno" column="deptno" />
    </resultMap>
    <select id="findEmpByIdGetMapper"  resultMap="empResultMap"
                                       parameterType="int">
        select * from emp where empno=#{empno}
    </select>
</mapper>
```

（3）在 mybatis-config.xml 配置文件中添加上述映射文件信息，代码如下。

```
<mappers>
     <mapper resource="com/ssmbook2020/ch7/dao/EmpMapper.xml" />
     <mapper resource="com/ssmbook2020/ch7/dao/EmpMapper2.xml" />
</mappers>
```

（4）不再创建 IEmpDao 的实现类，直接在 EmpController 控制器中利用接口动态代理机制实现接口中的方法。EmpController 中的代码如下。

```
@RequestMapping("/findEmpByIdGetMapper")
public ModelAndView findEmpByIdGetMapper(int empno) {
    SqlSession session=MyBatisUtil.getSession();
    IEmpDao2 empDao2=session.getMapper(IEmpDao2.class);
```

```
        Emp emp=empDao2.findEmpByIdGetMapper(empno);
        ModelAndView mv=new ModelAndView();
        mv.addObject("emp", emp);
        mv.setViewName("emp");
        return mv;
    }
```

其中代码 IEmpDao2 empDao2=session.getMapper(IEmpDao2.class)是利用了 SqlSession 对象的 getMaper 方法获得接口 IEmpDao2 的代理对象。然后与真实的接口实现类一样，正常调用接口的方法。

（5）在 WebContent 下创建 getMapper.jsp 文件，关键代码如下。

```
<body>
<form action="findEmpByIdGetMapper" method="post">
      员工编号：<input type="text" name="empno" />
      <input type="submit" value="查询员工" />
</form>
</body>
```

（6）运行测试，在浏览器地址栏中输入 http://localhost:8080/mybatis1/getMapper.jsp，在页面中输入一个员工编号，如图 7-20 所示。

图 7-20　查询员工信息

单击“查询员工”按钮，员工信息查询结果如图 7-21 所示。

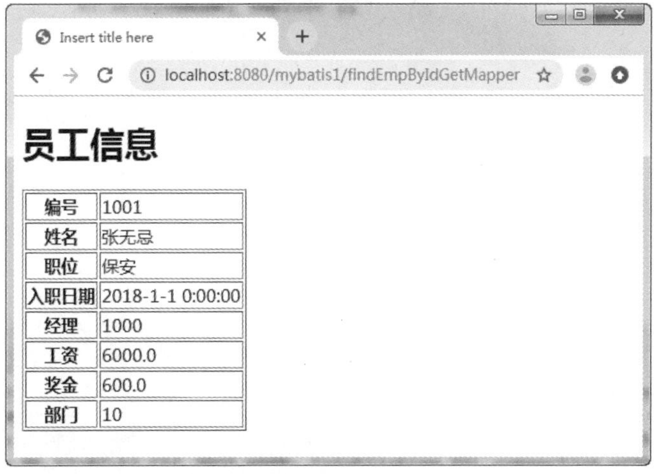

图 7-21　员工信息查询结果

习 题 7

1. MyBatis 对 JDBC 的优化和封装体现在哪些方面？
2. MyBatis 的优缺点是什么？
3. 简述 MyBatis 的工作流程。

上机练习 1

创建数据库 bookdb，创建表 book，字段为 bookid、bookname、author、price、content，添加若干测试数据。类比本章案例，使用 MyBatis 和 Spring MVC 进行 Web 界面的增、删、改、查操作。

第8章 MyBatis框架深入

本章学习内容

- MyBatis 动态查询；
- MyBatis 一对多查询；
- MyBatis 多对一查询；
- MyBatis 多对多查询；
- MyBatis 自连接查询；
- MyBatis 分页查询；
- MyBatis 缓存。

本章进一步学习 MyBatis 的动态查询、多表之间的关系、一对多查询、多对一查询、自连接查询、多对多查询、缓存等。

视频讲解

8.1 动态查询

动态查询在网购中很常见，大家在淘宝等网站中搜索商品时，搜索条件可多可少，每个人的搜索条件也都不尽相同，但最终都能查询出符合条件的商品。显然这些不同的搜索条件在后台动态地构造出了不同的 SQL 查询语句，从而能在数据库中查询到不同的商品。

动态 SQL 用于查询条件不确定的情况，根据用户提交的多种查询条件动态构造出 SQL 语句，这需要用到 MyBatis 的动态 SQL 标签，主要有<if>、<where>、<choose>、<foreach>、<sql>等。

8.1.1 <if>标签

语法：<if:test="布尔表达式">SQL 语句片段</if>

说明：该标签不能独立存在，必须嵌入<select>标签内部。如果布尔表达式的结果为 true，则<if>标签内部的"SQL 语句片段"就会被取用，与<select>标签中原来的 SQL 语句合并成一体；如果布尔表达式的结果为 false，则<if>标签内部的"SQL 语句片段"不会被取用，当作不存在。<select>标签内部可以有多个<if>标签，这样就可以组合成动态的 SQL 语句。具体做法见下面案例。

示例：动态查询员工信息。

实现步骤：

（1）创建项目 mybatis2，项目结构与上一章项目 mybatis1 相同，参考项目 mybatis1 搭建 Spring MVC 与 MyBatis 框架。在 com.ssmbook2020.ch8.dao 包下创建 IEmpDao 接口，添加 searchEmpsDynamic(Emp emp) 抽象方法。创建 EmpMapper.xml 映射文件，代码如下。

```
<?xml version="1.0" encoding="UTF-8"?>
<!DOCTYPE mapper
PUBLIC "-//mybatis.org//DTD Mapper 3.0//EN"
"http://mybatis.org/dtd/mybatis-3-mapper.dtd">
<mapper namespace="com.ssmbook2020.ch8.dao.IEmpDao">
    <!-- 配置结果映射 -->
```

```
        <resultMap id="empResultMap" type="com.ssmbook2020.ch8.entity.Emp">
            <id property="empno" column="empno" />
            <result property="empname" column="ename" />
            <result property="job" column="job" />
            <result property="manager" column="mgr" />
            <result property="hiredate" column="hiredate" />
            <result property="salary" column="sal" />
            <result property="commission" column="comm" />
            <result property="deptno" column="deptno" />
        </resultMap>
    <select id="searchEmpsDynamic" parameterType=" com.ssmbook2020.ch8.entity
.Emp"  resultMap="empResultMap">
            select * from emp where 1=1
            <if test="empno>0">
                and empno=#{empno}
            </if>
            <if test="empname!=null and empname!=''">
                and ename like '%' #{empname} '%'
            </if>
            <if test="job!=null and job!=''">
                and job like '%' #{job} '%'
            </if>
            <if test="manager>0">
                and mgr=#{manager}
            </if>
            <if test="hiredate!=null and hiredate!=''">
                and hiredate=#{hiredate}
            </if>
            <if test="salary>0">
                and sal>=#{salary}
            </if>
            <if test="commission>0">
                and comm=#{commission}
            </if>
            <if test="deptno>0">
                and deptno=#{deptno}
            </if>
        </select>
    </mapper>
```

这里的<select>语句块使用<if>标签构建动态 SQL 语句，使用 like '%' #{empname} '%' 语句构造模糊查询。

（2）在 com.ssmbook2020.ch7.dao 包下创建 EmpDaoImpl 接口实现类和实现接口的抽象方法，关键代码如下。

```
    public class EmpDaoImpl implements IEmpDao{
        @Override
```

```
        public List<Emp> searchEmpsDynamic(Emp emp) {
            SqlSession session=null;
            List<Emp> list=new ArrayList<Emp>();
            try {
                session=MyBatisUtil.getSession();
    list=session.selectList("com.ssmbook2020.ch8.dao.IEmpDao.searchEmpsDynamic",
emp);
                System.out.println(list.size());
            }catch (Exception e) {
                e.printStackTrace();
            }
            return list;
        }
    }
```

（3）在 com.ssmbook2020.ch8.controller 包下创建 EmpController 控制器，关键代码如下。

```
    @Controller
    public class EmpController {
        @RequestMapping("/searchEmpsDynamic")
        public ModelAndView searchEmpsDynamic(String empno,String empname,
                String job,String manager,String salary,String commission,String
                deptno) {
            EmpDaoImpl empDaoImpl=new EmpDaoImpl();
            Emp emp=new Emp();
            if(empno!=null) {
                emp.setEmpno(Integer.parseInt(empno));
            }else {
                emp.setEmpno(0);
            }
            if(manager!=null) {
                emp.setManager(Integer.parseInt(manager));
            }else {
                emp.setManager(0);
            }
            if(salary!=null) {
                emp.setSalary(Double.parseDouble(salary));
            }else {
                emp.setSalary(0.0);
            }
            if(commission!=null) {
                emp.setCommission(Double.parseDouble(commission));
            }else {
                emp.setCommission(0.0);
            }
            if(deptno!=null) {
```

```
            emp.setDeptno(Integer.parseInt(deptno));
        }else {
            emp.setDeptno(0);
        }
        emp.setEmpname(empname);
        emp.setJob(job);
        List<Emp> emps=empDaoImpl.searchEmpsDynamic(emp);
        ModelAndView mv=new ModelAndView();
        mv.addObject("emps", emps);
        mv.setViewName("emplist");
        return mv;
    }}
```

（4）在 WebContent 下创建 search.jsp 文件，将其作为查询界面，关键代码如下。

```
<body>
<form action="searchEmpsDynamic" method="post">
<table>
    <tr><th>编号</th> <td><input type="text" name="empno" value="0"/></td>
</tr>
    <tr><th>姓名</th><td><input type="text" name="empname"/></td></tr>
    <tr><th>职位</th><td><input type="text" name="job"/></td></tr>
    <tr><th>经理</th><td><input type="text" name="manager" value="0"/></td>
</tr>
    <tr><th>工资</th><td><input type="text" name="salary" value="0.0"/>
</td></tr>
    <tr><th>奖金</th><td><input type="text" name="commission" value="0.0"/>
</td></tr>
    <tr><th>部门</th><td><input type="text" name="deptno" value="0"/></td>
</tr>
    <tr><td colspan=2 align="center"><input type="submit" value="动态查询" />
</td></tr>
</table>
</form>
</body>
```

（5）在 WebContent/WEB-INF 目录下创建 jsp 目录，在 WebContent/WEB-INF/jsp 目录下创建 emplist.jsp，代码与项目 mybatis1 中的同名文件基本相同（可以复制过来），以此作为显示查询结果的视图。

（6）运行测试，在浏览器地址栏中输入 http://localhost:8080/mybatis2/search.jsp，出现查询界面。在界面中录入一些数据，如图 8-1 所示。

单击“动态查询”按钮，结果如图 8-2 所示。

可以多次尝试不同的查询条件进行查询。无论输入多少个条件，都能查询到满足条件的结果。这就是动态查询。

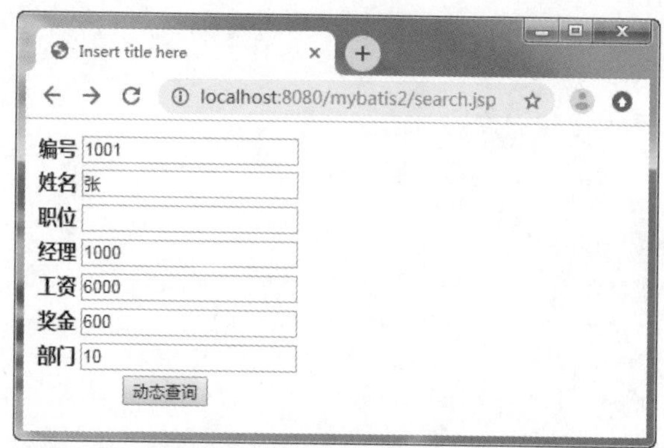

图 8-1　动态查询界面 1

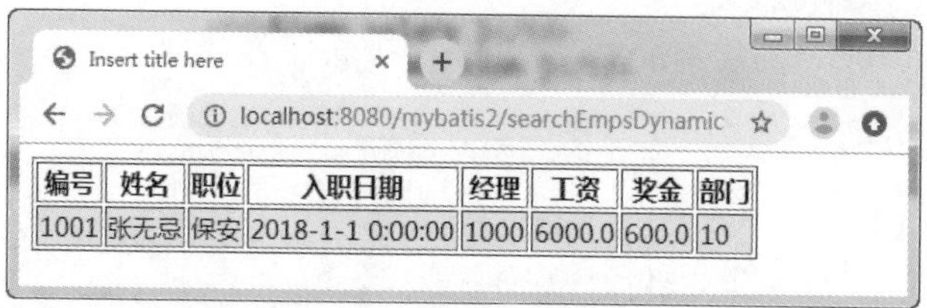

编号	姓名	职位	入职日期	经理	工资	奖金	部门
1001	张无忌	保安	2018-1-1 0:00:00	1000	6000.0	600.0	10

图 8-2　动态查询结果 1

8.1.2　<where>标签

在 8.1.1 节的案例中，select 语句的 where 后面使用了 1=1 似乎有点“奇怪”，这是为了使前面的 select 语句与后面的<if>标签里面的 SQL 语句能够顺利拼接而“发明”的一种“技巧”。还有一种办法，无须使用 1=1 也能顺利创建动态 SQL 语句，即使用<where>标签，具体用法见下面示例代码。

```
    <select id="searchEmpsDynamic" parameterType="com.ssmbook2020.ch8.entity
.Emp"  resultMap="empResultMap">
            select * from emp
        <where>
            <if test="empno>0">
                and empno=#{empno}
            </if>
            <if test="empname!=null and empname!=''">
                and ename like '%' #{empname} '%'
            </if>
            <if test="job!=null and job!=''">
                and job like '%'  #{job} '%'
            </if>
            <if test="manager>0">
                and mgr=#{manager}
            </if>
            <if test="hiredate!=null and hiredate!=''">
                and hiredate=#{hiredate}
            </if>
            <if test="salary>0">
```

```
            and sal=#{salary}
        </if>
        <if test="commission>0">
            and comm=#{commission}
        </if>
        <if test="deptno>0">
            and deptno=#{deptno}
        </if>
    </where>
</select>
```

将这些代码替换 EmpMapper.xml 映射文件中的原有<select>语句块，重新运行项目，测试结果相同。

8.1.3 <choose>标签

<choose>标签用来在多个条件中选择一个，完成类似 Java 中的 switch 的功能。<choose>标签包括<when>和<otherwise>子标签，一对<when></when>标签代表一个分支结构。<when>标签的 test 属性用于设置判断条件（布尔表达式），如果判断条件的结果为 true，则一对<when></when>标签内部的 SQL 语句会被取用，否则不取用。<choose>标签内部可以有多个<when>，代表多个条件分支，如果所有条件分支均不匹配，程序就执行到<otherwise>分支。

语法：

```
<choose>
    <when test="判断条件 1">SQL 语句块 1 </when>
    <when test="判断条件 2">SQL 语句块 2 </when>
    <when test="判断条件 3">SQL 语句块 3 </when>
    ...
    <otherwise>SQL 语句块 X</ otherwise >
</choose>
```

示例：如果员工姓名不为空，则按员工姓名查询；如果员工姓名为空但职位不为空，则按职位查询；如果以上条件都不满足，则查询全部员工。

实现步骤：

（1）在 IEmpDao 接口中添加 searchEmpsChoose(Emp emp) 方法，在 EmpMapper 映射文件中添加<select>代码块，实现多条件分支查询，代码如下。

```
<select id="searchEmpsChoose" parameterType="com.ssmbook2020.ch8.entity
.Emp"
    resultMap="empResultMap">
        select * from emp
        <where>
            <choose>
                <when test="empname!=null and empname!=''">
                    and ename like '%' #{empname} '%'
                </when>
```

```
                <when test="job!=null and job!=''">
                    and job like '%' #{job} '%'
                </when>
                <otherwise>
                    and 1=1
                </otherwise>
            </choose>
        </where>
    </select>
```

（2）在 EmpDaoImpl 接口实现类中重写 searchEmpsChoose(Emp emp)方法，代码如下。

```
    @Override
    public List<Emp> searchEmpsChoose(Emp emp) {
        SqlSession session=null;
        List<Emp> list=new ArrayList<Emp>();
        try {
            session=MyBatisUtil.getSession();
    list=session.selectList("com.ssmbook2020.ch8.dao.IEmpDao.searchEmpsChoose",
emp);
            System.out.println(list.size());
        }catch (Exception e) {
            e.printStackTrace();
        }
        return list;
    }
```

（3）在 EmpController 控制器中添加如下代码。

```
    @RequestMapping("/searchEmpsChoose")
    public ModelAndView searchEmpsChoose(Emp emp) {
        EmpDaoImpl empDaoImpl=new EmpDaoImpl();
        List<Emp> emps=empDaoImpl.searchEmpsChoose(emp);
        ModelAndView mv=new ModelAndView();
        mv.addObject("emps", emps);
        mv.setViewName("emplist");
        return mv;
    }
```

（4）在 WebContent 下创建 search2.jsp 文件，将其作为查询界面，关键代码如下。

```
    <body>
    <form action="searchEmpsChoose" method="post">
    <table>
        <tr><th>姓名</th><td><input type="text" name="empname"/></td></tr>
        <tr><th>职位</th><td><input type="text" name="job"/></td></tr>
        <tr><td colspan=2 align="center">
        <input type="submit" value="动态查询" />
        </td></tr>
```

```
    </table>
    </form>
    </body>
```

（5）运行测试，在浏览器地址栏中输入 http://localhost:8080/mybatis2/search2.jsp，出现如图 8-3 所示页面。

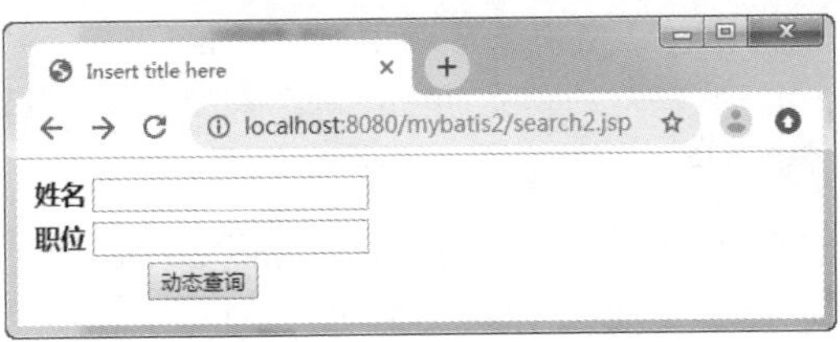

图 8-3　动态查询界面 2

姓名输入“张”，职位留空或任意输入，单击“动态查询”按钮，结果如图 8-4 所示。

编号	姓名	职位	入职日期	经理	工资	奖金	部门
1000	张三丰	老板	2000-1-1 0:00:00	0	50000.0	10000.0	10
1001	张无忌	保安	2018-1-1 0:00:00	1000	6000.0	600.0	10

图 8-4　动态查询结果 2

姓名留空，职位输入“工程师”，查询结果如图 8-5 所示。

编号	姓名	职位	入职日期	经理	工资	奖金	部门
1005	林冲	工程师	2018-3-1 0:00:00	1004	13000.0	4000.0	30

图 8-5　动态查询结果 3

姓名和职位都留空，查询结果如图 8-6 所示。由此可见，实现了示例要求的查询效果。

编号	姓名	职位	入职日期	经理	工资	奖金	部门
1000	张三丰	老板	2000-1-1 0:00:00	0	50000.0	10000.0	10
1001	张无忌	保安	2018-1-1 0:00:00	1000	6000.0	600.0	10
1002	李寻欢	项目经理	2017-2-1 0:00:00	1000	20000.0	5000.0	20
1003	黄飞鸿	程序员	2018-2-3 0:00:00	1002	10000.0	2000.0	20
1004	令狐冲	工程队长	2015-1-10 0:00:00	1000	18000.0	6000.0	30
1005	林冲	工程师	2018-3-1 0:00:00	1004	13000.0	4000.0	30
1006	武松	经理	2018-10-1 0:00:00	1004	20000.0	3000.0	30

图 8-6　动态查询结果 4

8.1.4 <foreach>标签

如果 MyBatis 的输入参数类型是数组或集合，则<select>语句块中可以用<foreach>子标签遍历这些数组或集合，从而构造所需的 SQL 语句。当输入参数类型是数组或集合时，一般无须在<select>标签中说明 parameterType 属性，MyBatis 会自动判断。

语法：

```
<foreach collection="array/list"open="起始符"close="结束符"item="迭代变量"
separator="分隔符"> #{迭代变量}</foreach>
```

说明：如果<foreach>遍历的是数组，则 collection="array"，这时 collection 的取值只能是"array"，不能是实际传入的数组参数名称。如果<foreach>遍历的是集合，则 collection="list"，这时 collection 的取值只能是"list"，不能是实际传入的集合参数名称。使用<foreach>标签，可以对传入的数组或集合进行遍历，构造成特定格式的 SQL 语句，再与其他 SQL 语句拼接成完整的 SQL 语句。

1．遍历数组

示例：查询编号为 1001、1003、1005 的员工，假设传入的参数是一个数组，其值为[1001,1003,1005]，映射文件中的<select>语句块设置如下。

```
<select id="searchEmpsForeach1" resultMap="empResultMap">
    select * from emp
    <if test="array!=null and array.length>0">
          where empno in
    <foreach collection="array" index="index" item="item" open="("
      separator = "," close=")">
            #{item}
    </foreach>
    </if>
</select>
```

可以看出，<foreach>标签将数组[1001,1003,1005]中的元素取出，构造出(1001,1003,1005)，再拼接到前面的 SQL 语句中，最终构造出下面的 SQL 语句。

```
select * from emp where empno in (1001,1003,1005)
```

在项目 mybatis2 中继续完善 DAO 层、控制层和视图，具体代码见教材源码。运行测试，在浏览器地址栏中输入 http://localhost:8080/mybatis2/search3.jsp，然后输入 1001、1003、1005，单击“查询”按钮，结果如图 8-7 所示。

图 8-7　动态查询结果 5

2．遍历集合

示例：查询员工信息，假设传入的参数为 emplist 集合，里面保存的是员工编号，映射文件中的<select>代码块如下所示。

```
<select id="searchEmpsForeach2"    resultMap="empResultMap">
     select * from emp
     <if test="list!=null and list.size>0">
          where empno in
          <foreach collection="list" open="(" item="item" close=")"
          separator=",">
                 #{item}
          </foreach>
     </if>
</select>
```

该代码将 emplist 集合中的每个对象元素取出，构造出类似(1001,1003,1005)格式的代码片段，再与前面的 SQL 语句拼接，形成如下所示的完整的 SQL 条件查询。

```
select * from emp where empno in (1001,1003,1005)
```

在项目 mybatis2 中继续完善 DAO 层、控制层和视图，具体代码见教材源码。运行测试，在浏览器地址栏中输入 http://localhost:8080/mybatis2/search4.jsp，然后输入 1001、1003、1005，单击“查询”按钮，结果如图 8-8 所示。

编号	姓名	职位	入职日期	经理	工资	奖金	部门
1001	张无忌	保安	2018-1-1 0:00:00	1000	6000.0	600.0	10
1003	黄飞鸿	程序员	2018-2-3 0:00:00	1002	10000.0	2000.0	20
1005	林冲	工程师	2018-3-1 0:00:00	1004	13000.0	4000.0	30

图 8-8　动态查询结果 6

8.1.5　<sql>标签

一个映射文件中可能有多个 SQL 语句块，有些 SQL 语句可能会被频繁重复使用，这时可以把重复的 SQL 定义为一个 SQL 片段，在需要重复使用的地方调用即可。简单来说，<sql>标签是用来定义 SQL 片段供其他标签复用的。定义 SQL 片段时，需要设置 id 属性进行标识。其他标签调用该 SQL 片段时，需要用到<include>标签，并用其 refid 属性指明要调用的 SQL 片段的 id 号。

示例：假设 select * from emp 语句被频繁复用，将其定义为代码片段，供其他标签调用，关键步骤如下。

（1）创建 SQL 片段，在映射文件中编写代码如下。

```
<sql id="selectEmp">
     select * from emp
```

```
</sql>
```

（2）在需要复用的地方调用该片段。这里假设 8.1.4 节的<foreach>遍历集合需要调用。

```
<select id="searchEmpsForeach2"    resultMap="empResultMap">
     <include refid="selectEmp"/>
     <if test="list!=null and list.size>0">
         where empno in
         <foreach collection="list" open="(" item="item" close=")"
          separator=",">
                 #{item}
          </foreach>
     </if>
</select>
```

上面代码中的第二行即为调用。

8.2 多表之间的关系

8.1 节的查询都是针对单个表的，如果有多个表该怎么操作呢？多表查询的结果如何封装为对象呢？下面讲解针对多个表的查询。为此，先要清楚两个表之间的各种关系。

两个表之间主要有三种关系:一对多、多对一和多对多。此外也有一对一的关系，但比较少见。

所谓一对多，就是“一方”表里的每一条记录都有“多方”表的多条记录与之对应。例如，部门表 dept 的每一条记录就是一个部门，而一个部门有多个员工，对应员工表 emp 里的多条记录。这里显然 dept 表是“一方”，emp 表是“多方”。

多对一，就是“多方”表里有多条记录对应“一方”表里的一条记录。例如，有多个员工属于同一个部门，则员工表 emp 里有多条记录对应部门表 dept 里的同一个记录。

多对多，就是一个表里的记录可能对应另一个表里的多条记录，一个表里的多条记录也可能对应另一个表里的一条记录。例如，一个员工可能参与多个项目，一个项目也可能有多个员工参加，员工表 emp 与项目表 project 之间就是多对多的关系。

一对一可以用多对一来处理，不再单独介绍。

下面对这些关系进行详细讲解。

视频讲解

8.3 一对多查询

一对多查询需要在查询“一方”数据的同时把“多方”数据也查询出来。“一方”的每一条记录都封装为一个对象，“一方”的对象需要额外定义一个“多方”类型的泛型集合属性，将“多方”的数据查询出来并封装为泛型集合后“存储”到这个属性中。“多方”对象也可以定义一个“一方”类型的属性。

示例：输入一个部门编号，查询该部门信息和该部门的所有员工信息。

实现步骤：

（1）在项目 mybatis2 中添加实体类 Dept，关键代码如下。

```
public class Dept {
    private int deptno;
    private String dname;
    private String loc;
    private List<Emp> emps;
    //省略其他方法
}
```

除数据表中对应列的属性外，还有一个泛型集合属性 emps，用于存储该部门的所有员工信息。修改 Emp 类，添加一个 Dept 类型的属性 dept，代表该员工所在的部门信息，同时补充 getter、setter 方法。新添加的代码如下。

```
private Dept dept;
public Dept getDept() {
     return dept;
}
public void setDept(Dept dept) {
     this.dept = dept;
}
```

（2）创建 IDeptDao 接口，添加 findEmpsByDept(int deptno)方法，创建 DeptMapper.xml 映射文件，代码如下。

```
<?xml version="1.0" encoding="UTF-8"?>
<!DOCTYPE mapper PUBLIC "-//mybatis.org//DTD Mapper 3.0//EN"
"http://mybatis.org/dtd/mybatis-3-mapper.dtd">
<mapper namespace="com.ssmbook2020.ch8.dao.IDeptDao">
    <resultMap id="deptResultMap" type="com.ssmbook2020.ch8.entity.Dept">
        <id property="deptno" column="deptno" />
        <result property="dname" column="dname" />
        <result property="loc" column="loc" />
        <collection property="emps" ofType="com.ssmbook2020.ch8.entity
.Emp">
            <id property="empno" column="empno" />
            <result property="empname" column="ename" />
            <result property="job" column="job" />
            <result property="manager" column="mgr" />
            <result property="hiredate" column="hiredate" />
            <result property="salary" column="sal" />
            <result property="commission" column="comm" />
        </collection>
    </resultMap>
```

```
        <select id="findEmpsByDept" resultMap="deptResultMap" parameterType=
"int">
            select empno,ename,job,mgr,hiredate,sal,comm,dept.deptno,dname,loc
            from emp,dept where  emp.deptno=dept.deptno and dept.deptno=
#{deptno}
        </select>
    </mapper>
```

这里的关键技术：映射文件中用到的一对多查询是两表连接查询，一次性查询两张表的数据，然后用结果映射 ResultMap 把与 dept 表有关的查询结果映射到 Dept 对象中。结果映射中使用<collection>标签把与 emp 表有关的查询结果封装为 Emp 对象，再映射到 Dept 对象的 emps 属性中。

此外要在 MyBatis 配置文件 mybatis-conf.xml 中指定该映射文件：

```
    <mapper resource="com/ssmbook2020/ch8/dao/DeptMapper.xml" />
```

（3）创建接口实现类 DeptDaoImpl，关键代码如下。

```
    public class DeptDaoImpl implements IDeptDao{
        @Override
        public Dept findEmpsByDept(int deptno) {
            SqlSession session=null;
            Dept dept=null;
            List<Emp> emps=new ArrayList<Emp>();
            try {
                session=MyBatisUtil.getSession();
    dept=session.selectOne("com.ssmbook2020.ch8.dao.IDeptDao.findEmpsByDept"
,deptno);
            }catch (Exception e) {
                e.printStackTrace();
            }
            return dept;
        }
    }
```

（4）创建 DeptController 控制器，关键代码如下。

```
    @Controller
    public class DeptController {
        @RequestMapping("/findEmpsByDept")
        public ModelAndView findEmpsByDept(int deptno) {
            DeptDaoImpl deptDao=new DeptDaoImpl();
            Dept dept=deptDao.findEmpsByDept(deptno);
            ModelAndView mv=new ModelAndView();
            mv.addObject("dept", dept);
            mv.setViewName("forward:search5.jsp");
            return mv;
```

```
    }
}
```

（5）在 WebContent 下创建 search5.jsp 页面，关键代码如下。

```
<body>
<h1>查询部门信息 1</h1>
<form action="findEmpsByDept" method="post">
部门编号：<input type="text" name="deptno" />
<input type="submit" value="查询" />
</form>
<h1>查询部门信息 2</h1>
<form action="findEmpsByDept2" method="post">
部门编号：<input type="text" name="deptno" />
<input type="submit" value="查询" />
</form>
<c:if test="${not empty dept }">
<hr/>
部门编号：${dept.deptno }<br/>
部门名称：${dept.dname }<br/>
部门地址：${dept.loc }<br/>
所属员工：<br/>
<table border="1">
        <tr>
            <th>员工编号</th>
            <th>姓名</th>
            <th>职位</th>
            <th>入职日期</th>
            <th>经理</th>
            <th>工资</th>
            <th>奖金</th>
        </tr>
        <c:forEach items="${dept.emps }" var="emp" varStatus="vs">
            <tr ${vs.count%2==1 ? "style='background-color:yellow'" :
            "style='background-color:white'" }>
                <td>${emp.empno}</td>
                <td>${emp.empname }</td>
                <td>${emp.job }</td>
                <td>${emp.hiredate.toLocaleString() }</td>
                <td>${emp.manager }</td>
                <td>${emp.salary }</td>
                <td>${emp.commission }</td>
            </tr>
        </c:forEach>
    </table>
    </c:if>
</body>
```

（6）运行测试，在浏览器地址栏中输入 http://localhost:8080/mybatis2/search5.jsp，出现如图 8-9 所示的查询页面。在“查询部门信息 1”下输入 30，结果如图 8-10 所示。

图 8-9　部门信息查询界面

图 8-10　部门信息查询结果 1

由此可见，在查询部门信息的同时，其所属员工信息也被查询出来了，实现了一对多查询。

除使用两表连接查询外，还可以在映射文件中使用两次查询的技术，分别单独查询 dept 表和 emp 表。具体见下面的有关步骤。

（1）在 IDeptDao 接口中添加 findEmpsByDept2(int deptno) 方法，在 DeptMapper.xml 映射文件中添加代码如下。

```
    <select id="findEmpsByDept2" resultMap="deptResultMap2" parameterType=
"int">
        select * from dept where deptno=${deptno}
    </select>
```

```
        <resultMap id="deptResultMap2" type="com.ssmbook2020.ch8.entity.Dept">
            <id property="deptno" column="deptno" />
            <result property="dname" column="dname" />
            <result property="loc" column="loc" />
            <!-- 关联属性的映射关系集合的数据来自指定的select查询,该select查询的动态参
数来自 column 指定的字段值 -->
            <collection property="emps" ofType="com.ssmbook2020.ch8.entity.Emp"
            select="findEmpsByDept3" column="deptno">
          </collection>
    </resultMap>

    <select id="findEmpsByDept3" resultMap="deptResultMap3" parameterType=
"int">
        select * from emp where deptno=${deptno}
    </select>
    <resultMap id="deptResultMap3" type="com.ssmbook2020.ch8.entity.Emp">
        <id property="empno" column="empno" />
        <result property="empname" column="ename" />
        <result property="job" column="job" />
        <result property="manager" column="mgr" />
        <result property="hiredate" column="hiredate" />
        <result property="salary" column="sal" />
        <result property="commission" column="comm" />
    </resultMap>
```

这里先根据 deptno 查询 dept 表，查询的结果用 id 为 deptResultMap2 的 ResultMap 封装成 Dept 对象，ResultMap 中使用<collection>标签封装 Dept 对象的 emps 属性，同时指定使用 id 号为 findEmpsByDept3 的<select>语句块来查询 emps 的有关数据，并且指定该查询的动态参数来自 column 指定的值，然后查询 emp 表并指定 id 号为 deptResultMap3 的 ResultMap 封装成 Emp 对象。

（2）在接口实现类 DeptDaoImpl 中添加如下方法。

```
    @Override
    public Dept findEmpsByDept2(int deptno) {
        SqlSession session=null;
        Dept dept=null;
        try {
            session=MyBatisUtil.getSession();
        dept=session.selectOne("com.ssmbook2020.ch8.dao.IDeptDao.
findEmpsByDept2",deptno);
        }catch (Exception e) {
            e.printStackTrace();
        }
        return dept;
    }
```

（3）在 DeptController 控制器中添加下述方法。

```
@RequestMapping("/findEmpsByDept2")
public ModelAndView findEmpsByDept2(int deptno) {
    DeptDaoImpl deptDao=new DeptDaoImpl();
    Dept dept=deptDao.findEmpsByDept2(deptno);
    ModelAndView mv=new ModelAndView();
    mv.addObject("dept", dept);
    mv.setViewName("forward:search5");
    return mv;
}
```

（4）运行测试，在浏览器地址栏中输入 http://localhost:8080/mybatis2/search5.jsp，出现查询页面，在“查询部门信息 2”下输入 30，结果如图 8-11 所示。

图 8-11　部门信息查询结果 2

可见两种方式都可以实现一对多查询，但推荐使用第一种。

8.4 多对一查询

多对一查询要在“多方”的实体类中设置“一方”类型的属性。例如，员工 Emp 是“多方”，Dept 是“一方”，需要在 Emp 中设置一个 Dept 类型的属性 dept，代表员工所在的部门信息。

关键技术：SQL 查询采用两表连接查询，同时查询两个表的信息，用代表“多方”的结果映射进行封装，在“多方”的映射文件中用<association>标签关联“一方”的信息。具体做法见下面示例。

示例：查询一个员工的信息，同时查询该员工所在部门的信息。

实现步骤：

（1）在 IEmpDao 接口中添加 findEmpByIdWithDept(int empno)方法，在 EmpMapper.xml 映射文件中添加如下代码。

```
    <select id="findEmpByIdWithDept" resultMap="empResultMap2"
    parameterType="int">
          select e.*,d.* from emp e,dept d where e.empno=#{empno}
    </select>
    <resultMap id="empResultMap2" type="com.ssmbook2020.ch8.entity.Emp">
         <id property="empno" column="empno" />
         <result property="empname" column="ename" />
         <result property="job" column="job" />
         <result property="manager" column="mgr" />
         <result property="hiredate" column="hiredate" />
         <result property="salary" column="sal" />
         <result property="commission" column="comm" />
         <association property="dept" javaType="com.ssmbook2020.ch8.entity
.Dept">
              <id property="deptno" column="deptno" />
              <result property="dname" column="dname" />
              <result property="loc" column="loc" />
         </association>
    </resultMap>
```

（2）在接口实现类 DeptDaoImpl 中添加如下方法。

```
    @Override
    public Emp findEmpByIdWithDept(int empno) {
        SqlSession session=null;
        Emp emp=null;
        try {
            session=MyBatisUtil.getSession();
    emp=session.selectOne("com.ssmbook2020.ch8.dao.IEmpDao.findEmpByIdWithDept",
empno);
        }catch (Exception e) {
            e.printStackTrace();
        }
        return emp;
    }
```

（3）在 DeptController 控制器中添加如下方法。

```
    @RequestMapping("/findEmpByIdWithDept")
    public ModelAndView findEmpByIdWithDept(int empno) {
        EmpDaoImpl empDaoImpl=new EmpDaoImpl();
        Emp emp=empDaoImpl.findEmpByIdWithDept(empno);
        ModelAndView mv=new ModelAndView();
```

```
        mv.addObject("emp", emp);
        mv.setViewName("forward:search6.jsp");
        return mv;
    }
```

（4）在 WebContent 下创建 search6.jsp 页面，关键代码如下。

```
<body>
<h1>查询员工信息 1</h1>
<form action="findEmpByIdWithDept" method="post">
员工编号：<input type="text" name="empno" />
<input type="submit" value="查询" />
</form>
<h1>查询员工信息 2</h1>
<form action="findEmpByIdWithDept2" method="post">
员工编号：<input type="text" name="empno" />
<input type="submit" value="查询" />
</form>
<c:if test="${not empty emp }">
<hr/>
<h1>员工信息</h1>
<table border="1">
    <tr><th>编号</th><td>${emp.empno }</td></tr>
    <tr><th>姓名</th><td>${emp.empname}</td>    </tr>
    <tr><th>职位</th><td>${emp.job }</td>    </tr>
    <tr><th>入职日期</th><td>${emp.hiredate.toLocaleString() }</td> </tr>
    <tr><th>经理</th><td>${emp.manager }</td>    </tr>
    <tr><th>工资</th><td>${emp.salary }</td>    </tr>
    <tr><th>奖金</th><td>${emp.commission }</td>    </tr>
</table>
所在部门信息：<br/>
部门编号：${emp.dept.deptno }
部门名称：${emp.dept.dname }
部门地址：${emp.dept.loc }
</c:if>
</body>
```

（5）运行测试，在浏览器地址栏中输入 http://localhost:8080/mybatis2/search6.jsp，出现如图 8-12 所示的页面。在“查询员工信息 1”下面的文本框中输入 1002，单击“查询”按钮，结果如图 8-13 所示。

由此可见，在查询员工信息的同时，其部门信息也被查询出来了，实现了多对一查询。

还有一种方法也能实现多对一查询，就是两个表单独查询并分别封装，具体如下。

（1）在 IEmpDao 接口中添加 findEmpByIdWithDept2(int empno)方法，在 EmpMapper.xml 映射文件中添加代码如下。

图 8-12　员工信息查询界面 1

图 8-13　员工信息查询结果 1

```
<select id="findEmpByIdWithDept2" resultMap="empResultMap3" parameterType
                                        ="int">
    select * from emp  where empno=#{empno}
</select>
<resultMap id="empResultMap3" type="com.ssmbook2020.ch8.entity.Emp">
    <id property="empno" column="empno" />
    <result property="empname" column="ename" />
    <result property="job" column="job" />
    <result property="hiredate" column="hiredate" />
    <result property="salary" column="sal" />
    <result property="commission" column="comm" />
```

```
        <association property="dept" javaType="com.ssmbook2020.ch8.entity
.Dept"
                select="findDeptByDeptno" column="deptno">
            </association>
    </resultMap>
    <select id="findDeptByDeptno" resultType="com.ssmbook2020.ch8.entity.Dept"
                                parameterType="int">
        select * from dept  where deptno=#{deptno}
    </select>
```

首先查询 emp 表，根据 id 为 empResultMap3 的 ResultMap 封装为 Emp，在 ResultMap 中用<association>标签封装 dept 属性，并指定用 id 号为 findDeptByDeptno 的 SQL 语句块获取数据，使用 column 指定的 deptno 作为动态参数。然后执行 findDeptByDeptno 语句块，查询 dept 表并封装为对象。

（2）在接口实现类 EmpDapImpl 中添加如下方法。

```
    @Override
    public Emp findEmpByIdWithDept2(int empno) {
        SqlSession session=null;
        Emp emp=null;
        try {
            session=MyBatisUtil.getSession();
    emp=session.selectOne("com.ssmbook2020.ch8.dao.IEmpDao.findEmpByIdWithDept2",
empno);
        }catch (Exception e) {
            e.printStackTrace();
        }
        return emp;
    }
```

（3）在 EmpController 控制器中添加如下方法。

```
    @RequestMapping("/findEmpByIdWithDept2")
    public ModelAndView findEmpByIdWithDept2(int empno) {
        EmpDaoImpl empDaoImpl=new EmpDaoImpl();
        Emp emp=empDaoImpl.findEmpByIdWithDept2(empno);
        ModelAndView mv=new ModelAndView();
        mv.addObject("emp", emp);
        mv.setViewName("forward:search6.jsp");
        return mv;
    }
```

（4）运行测试，在浏览器地址栏中输入 http://localhost:8080/mybatis2/search6.jsp。在“查询员工信息 2”下面的文本框中输入 1002，单击“查询”按钮，结果如图 8-14 所示。

可见这种方法也可以实现多对一查询，但推荐使用第一种方法。

图 8-14　员工信息查询结果 2

8.5 自连接查询

所谓自连接，就是自己连接自己，比如员工表中有一个字段是经理编号，但经理本身也是员工。如果要查询某个员工的经理，就需要自己连接自己，否则只能查到经理的编号，而不能知道该经理的详细信息。

如果要查找某个员工的经理，可以用多对一的方式实现自连接；如果要查找某个经理所属的员工，则可以用一对多的方式实现自连接。

8.5.1 以多对一的方式实现自连接

示例：查询某个员工的信息及其经理信息。

分析：利用多对一的查询方式，即将员工作为“多方”、经理作为“一方”进行查询。

实现步骤：

（1）在 mybatis2 项目中，修改实体类 Emp，添加一个 Emp 类型的属性 leader，用于存储其上级领导，同时设置其 getter、setter 方法。

```
private Emp leader;
public Emp getLeader() {
    return leader;
}
public void setLeader(Emp leader) {
```

```
        this.leader = leader;
    }
```

（2）在 IEmpDao 接口中添加 findEmpLeaderById(int id) 方法。

```
    public Emp findEmpLeaderById (int empno);
```

（3）在 EmpMapper.xml 映射文件中添加如下代码。

```
    <select id="findEmpLeaderById" parameterType="int"
    resultMap="empResultMap4">
            select * from emp where empno=#{empno}
        </select>
        <resultMap id="empResultMap4" type="com.ssmbook2020.ch8.entity.Emp">
            <id property="empno" column="empno" />
            <result property="empname" column="ename" />
            <result property="job" column="job" />
            <result property="hiredate" column="hiredate" />
            <result property="salary" column="sal" />
            <result property="commission" column="comm" />
            <result property="deptno" column="deptno" />
            <association property="leader" javaType="com.ssmbook2020.ch8.entity
             .Emp"
             select="findEmpLeaderById" column="mgr">
             </association>
        </resultMap>
```

（4）在接口实现类 EmpDaoImpl 中添加如下代码。

```
    @Override
    public Emp findEmpLeaderById(int empno) {
        SqlSession session=null;
        Emp emp=null;
        try {
            session=MyBatisUtil.getSession();
        emp=session.selectOne("com.ssmbook2020.ch8.dao.IEmpDao.
findEmpLeaderById", empno);
        }catch (Exception e) {
            e.printStackTrace();
        }
        return emp;
    }
```

（5）在控制器 EmpController 中添加如下方法。

```
    @RequestMapping("/findEmpLeaderById")
    public ModelAndView findEmpLeaderById(int empno) {
        EmpDaoImpl empDaoImpl=new EmpDaoImpl();
        Emp emp=empDaoImpl.findEmpLeaderById(empno);
```

```
        ModelAndView mv=new ModelAndView();
        mv.addObject("emp", emp);
        mv.setViewName("forward:search7.jsp");
        return mv;
    }
```

（6）在 WebContent 下创建 search7.jsp，关键代码如下。

```
<body>
<h1>查询员工信息 1</h1>
<form action="findEmpLeaderById" method="post">
员工编号：<input type="text" name="empno" />
<input type="submit" value="查询" />
</form>
<c:if test="${not empty emp }">
<hr/>
<h1>员工信息</h1>
<table border="1">
    <tr><th>编号</th><td>${emp.empno }</td>    </tr>
    <tr><th>姓名</th><td>${emp.empname}</td>    </tr>
    <tr><th>职位</th><td>${emp.job }</td>    </tr>
    <tr><th>入职日期</th><td>${emp.hiredate.toLocaleString() }</td> </tr>
    <tr><th>经理</th><td>${emp.manager }</td>    </tr>
    <tr><th>工资</th><td>${emp.salary }</td>    </tr>
    <tr><th>奖金</th><td>${emp.commission }</td>    </tr>
    <tr><th>部门</th><td>${emp.deptno }</td>    </tr>
</table>
  员工的经理信息：<br/>
<table border="1">
    <tr><th>编号</th><td>${emp.leader.empno }</td>    </tr>
    <tr><th>姓名</th><td>${emp.leader.empname}</td>    </tr>
    <tr><th>职位</th><td>${emp.leader.job }</td>    </tr>
    <tr><th>入职日期</th><td>${emp.leader.hiredate.toLocaleString()}</td>
</tr>
    <tr><th>经理</th><td>${emp.leader.manager }</td>    </tr>
    <tr><th>工资</th><td>${emp.leader.salary }</td>    </tr>
    <tr><th>奖金</th><td>${emp.leader.commission }</td>    </tr>
    <tr><th>部门</th><td>${emp.leader.deptno }</td>    </tr>
</table>
</c:if>
</body>
</html>
```

（7）运行测试，在浏览器地址栏中输入 http://localhost:8080/mybatis2/search7.jsp。在员工编号右侧的文本框中输入 1002，单击“查询”按钮，结果如图 8-15 所示。

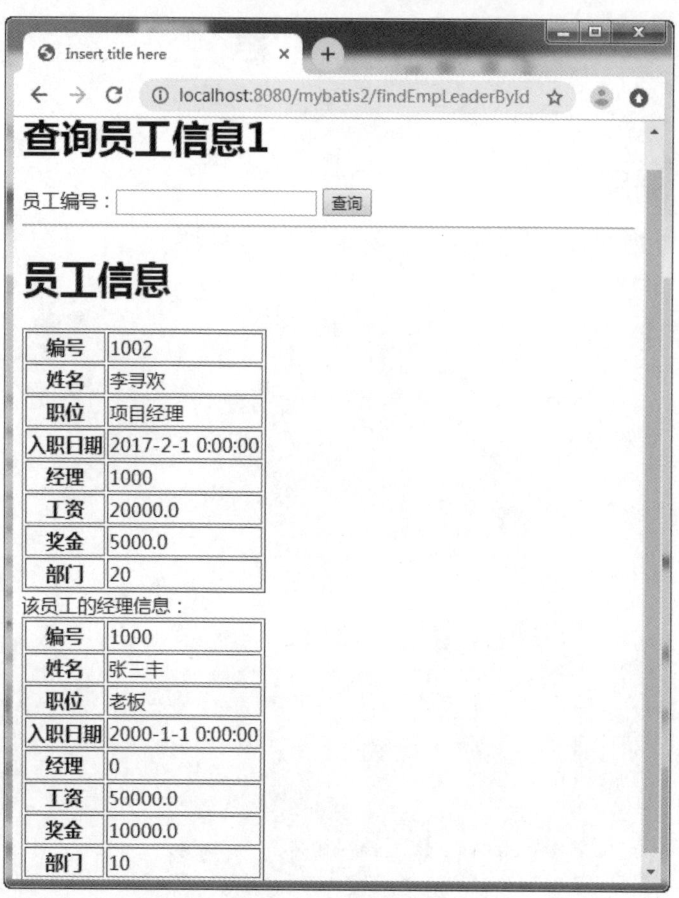

Insert title here
localhost:8080/mybatis2/findEmpLeaderById
查询员工信息1
员工编号：
查询
员工信息
编号 1002
姓名 李寻欢
职位 项目经理
入职日期 2017-2-1 0:00:00
经理 1000
工资 20000.0
奖金 5000.0
部门 20
该员工的经理信息：
编号 1000
姓名 张三丰
职位 老板
入职日期 2000-1-1 0:00:00
经理 0
工资 50000.0
奖金 10000.0
部门 10

图 8-15　某员工及其经理信息查询结果

8.5.2　以一对多的方式实现自连接

示例：查询某个经理信息及其下属员工信息。

分析：可用一对多的方式来实现，员工（经理）当作“一方”，员工（下属）当作“多方”。

实现步骤：

（1）修改实体类 Emp，添加 List<Emp>类型的 subemps 属性，用于存储其下属员工。添加的代码如下。

```
private List<Emp> subemps; //下属员工
public List<Emp> getSubemps() {
    return subemps;
}
public void setSubemps(List<Emp> subemps) {
    this.subemps = subemps;
}
```

（2）在 IEmpDao 接口中添加如下方法。

```
public Emp findEmpLeaderById2(int empno);
```

（3）在 EmpMapper.xml 映射文件中添加如下代码。

```
<select id="findEmpLeaderById2" parameterType="int"
resultMap="empResultMap5">
    select * from emp where empno=#{empno}
</select>
<resultMap id="empResultMap5" type="com.ssmbook2020.ch8.entity.Emp">
    <id property="empno" column="empno" />
    <result property="empname" column="ename" />
    <result property="job" column="job" />
    <result property="hiredate" column="hiredate" />
    <result property="salary" column="sal" />
    <result property="commission" column="comm" />
    <result property="deptno" column="deptno" />
    <result property="manager" column="mgr" />
    <collection property="subemps" ofType="com.ssmbook2020.ch8.entity.Emp"
          select="selectSubemps" column="empno">
    </collection>
</resultMap>
<select id="selectSubemps" resultMap="empResultMap">
    select * from emp where mgr=#{empno}
</select>
```

（4）在接口实现类 EmpDaoImpl.java 中添加 findEmpLeaderById2(int empno) 方法。

```
@Override
public Emp findEmpLeaderById2(int empno) {
    SqlSession session=null;
    Emp emp=null;
    try {
        session=MyBatisUtil.getSession();
    emp=session.selectOne("com.ssmbook2020.ch8.dao.IEmpDao.
findEmpLeaderById2",empno);
    }catch (Exception e) {
        e.printStackTrace();
    }
    return emp;
}
```

（5）在 EmpController 控制器中添加如下方法。

```
@RequestMapping("/findEmpLeaderById2")
public ModelAndView findEmpLeaderById2(int empno) {
    EmpDaoImpl empDaoImpl=new EmpDaoImpl();
    Emp emp=empDaoImpl.findEmpLeaderById2(empno);
    ModelAndView mv=new ModelAndView();
    mv.addObject("emp", emp);
    mv.setViewName("forward:search8.jsp");
```

```
    return mv;
}
```

（6）在 WebContent 下创建 search8.jsp 页面，关键代码如下。

```
<body>
<h1>查询员工信息 2</h1>
<form action="findEmpLeaderById2" method="post">
员工编号：<input type="text" name="empno" />
<input type="submit" value="查询" />
</form>
<c:if test="${not empty emp }">
<hr/>
<h1>员工信息</h1>
<table border="1">
    <tr><th>编号</th><td>${emp.empno }</td>    </tr>
    <tr><th>姓名</th><td>${emp.empname}</td>    </tr>
    <tr><th>职位</th><td>${emp.job }</td>    </tr>
    <tr><th>入职日期</th><td>${emp.hiredate.toLocaleString() }</td>  </tr>
    <tr><th>经理</th><td>${emp.manager }</td>    </tr>
    <tr><th>工资</th><td>${emp.salary }</td>    </tr>
    <tr><th>奖金</th><td>${emp.commission }</td>    </tr>
    <tr><th>部门</th><td>${emp.deptno }</td>    </tr>
</table>
  下属员工信息：<br/>
<table border="1">
        <tr>
            <th>编号</th><th>姓名</th><th>职位</th>    <th>入职日期</th>
            <th>经理</th><th>工资</th><th>奖金</th>    <th>部门</th>
        </tr>
        <c:forEach items="${emp.subemps }" var="emp" varStatus="vs">
            <tr ${vs.count%2==1 ? "style='background-color:yellow'" :
                                 "style='background-color:white'" }>
                <td>${emp.empno}</td>
                <td>${emp.empname }</td>
                <td>${emp.job }</td>
                <td>${emp.hiredate.toLocaleString() }</td>
                <td>${emp.manager }</td>
                <td>${emp.salary }</td>
                <td>${emp.commission }</td>
                <td>${emp.deptno }</td>
            </tr>
        </c:forEach>
    </table>
</c:if>
</body>
```

（7）运行测试，在浏览器地址栏中输入 http://localhost:8080/mybatis2/search8.jsp，结果

如图 8-16 所示。在员工编号右侧文本框中输入 1004，单击“查询”按钮，结果如图 8-17 所示。

图 8-16 员工信息查询界面 2

图 8-17 某经理及其下属员工信息查询结果

8.6 多对多查询

多对多查询通常需要一个中间表。例如，员工表 emp 与项目表 project，一个员工可以参与多个项目，一个项目也可以有多个员工参与，这两个表显然是多对多的关系，但这两个表之间并无外键，只凭它们无法建立关系。为此，需要员工参与项目表 projectemp 这个中间表，该表有员工编号、项目编号和工作内容等字段，记录员工参与项目的情况。emp

表与中间表 projectemp 是一对多的关系，表示一个员工可以参与多个项目，project 表与中间表 projectemp 也是一对多的关系，表示一个项目可以有多个员工参与。多对多实际上由两个一对多组成，分别实现两个一对多即可实现多对多。

示例：查询一个员工参与项目的情况和一个项目的人员参与情况。

实现步骤：

（1）在 MySQL 数据库中创建 project 表，用于存储项目信息，projectemp 表用于存储员工参与项目的信息，各录入若干测试数据，如图 8-18 和图 8-19 所示。

projectid	projectname
1	员工信息管理系统
2	网上商城
3	人事管理系统

图 8-18　project 表

id	projectid	empno	content
1	1	1001	前端
2	1	1002	编码
3	1	1003	数据库
4	2	1002	后台
5	2	1003	测试
6	2	1004	维护
7	2	1005	咨询
8	3	1002	设计
9	3	1001	网络

图 8-19　projectemp 表

（2）在项目 mybatis2 中创建实体类 Project，其中需要添加一个 List<Emp>类型的属性 emps，用于存储参与此项目的员工集合。修改 Emp 类，添加一个 List<Project>类型的属性 projects，用于存储该员工参与的项目集合，同时添加相关的 getter、setter 方法。Project 类的关键代码如下。

```
public class Project {
    private int projectid;
    private String projectname;
    private List<Emp> emps;      //参与该项目的员工集合
    private String content;      //员工参与其中一个项目的工作内容
    //省略 getter/setter 方法
}
```

Emp 类中新添加的属性如下。

```
private List<Project> projects;//该员工参与的项目集合
public List<Project> getProjects() {
    return projects;
}
public void setProjects(List<Project> projects) {
    this.projects = projects;
}
private String content;//项目中某一位员工参与的工作内容
public String getContent() {
    return content;
}
public void setContent(String content) {
```

```
        this.content = content;
    }
```

注意：content 不是重点，可对此忽略（下同）。

（3）创建 IProjectDao 接口，添加 findProjectEmps(int projected)方法，通过项目 projected 查询项目信息及其参与人员的信息。在 IEmpDao 接口中添加 findEmpProjects(int empno) 方法，通过员工的 empno 查询员工信息及其参与的项目信息。

（4）创建 ProjectMapper 映射文件，代码如下。

```
<?xml version="1.0" encoding="UTF-8"?>
<!DOCTYPE mapper
PUBLIC "-//mybatis.org//DTD Mapper 3.0//EN"
"http://mybatis.org/dtd/mybatis-3-mapper.dtd">
<mapper namespace="com.ssmbook2020.ch8.dao.IProjectDao">
    <select id="findProjectEmps" resultMap="projectResultMap"
    parameterType="int">
        select e.*,p.*,pe.content from emp e,projectemp pe,project p
        where  e.empno=pe.empno and p.projectid=pe.projectid and
        p.projectid=#{projectid}
    </select>
    <resultMap id="projectResultMap" type="com.ssmbook2020.ch8.entity.Project">
        <id property="projectid" column="projectid" />
        <result property="projectname" column="projectname" />
        <collection property="emps" ofType="com.ssmbook2020.ch8.entity.Emp">
            <id property="empno" column="empno" />
            <result property="empname" column="ename" />
            <result property="job" column="job" />
            <result property="manager" column="mgr" />
            <result property="hiredate" column="hiredate" />
            <result property="salary" column="sal" />
            <result property="commission" column="comm" />
            <result property="deptno" column="deptno" />
            <result property="content" column="content" />
        </collection>
    </resultMap>
</mapper>
```

注意要在配置文件 mybatis-config.xml 中指定该映射文件，代码如下。

```
<mappers>
     <mapper resource="com/ssmbook2020/ch8/dao/EmpMapper.xml" />
     <mapper resource="com/ssmbook2020/ch8/dao/DeptMapper.xml" />
     <mapper resource="com/ssmbook2020/ch8/dao/ProjectMapper.xml" />
</mappers>
```

（5）在 EmpMapper.xml 映射文件中添加如下代码。

```
    <select id="findEmpProjects" resultMap="empResultMap6" parameterType="int">
         select e.*,p.*,pe.content from emp e,projectemp pe,project p where
          e.empno=pe.empno and p.projectid=pe.projectid and e.empno=#{empno}
    </select>
    <resultMap id="empResultMap6" type="com.ssmbook2020.ch8.entity.Emp">
         <id property="empno" column="empno" />
         <result property="empname" column="ename" />
         <result property="job" column="job" />
         <result property="hiredate" column="hiredate" />
         <result property="salary" column="sal" />
         <result property="commission" column="comm" />
         <result property="manager" column="mgr" />
         <result property="deptno" column="deptno" />
         <collection property="projects"
              ofType="com.ssmbook2020.ch8.entity.Project" >
              <id property="projectid" column="projectid" />
              <result property="projectname" column="projectname" />
              <result property="content" column="content" />
         </collection>
    </resultMap>
```

（6）创建 IProjectDao 接口的 ProjectDaoImpl 实现类，添加如下代码。

```
    @Override
    public Project findProjectEmps(int projectid) {
        SqlSession session=null;
        Project project=null;
        try {
            session=MyBatisUtil.getSession();
            project=session.selectOne("com.ssmbook2020.ch8.dao.IProjectDao.
findProjectEmps",projectid);
        }catch (Exception e) {
             e.printStackTrace();
        }
        return project;
    }
```

（7）在 IEmpDao 接口的 EmpDaoImpl 实现类中添加如下方法。

```
    @Override
    public Emp findEmpProjects(int empno) {
        SqlSession session=null;
        Emp emp=null;
        try {
            session=MyBatisUtil.getSession();
            emp=session.selectOne("com.ssmbook2020.ch8.dao.IProjectDao.
```

```
findEmpProjects",empno);
        }catch (Exception e) {
             e.printStackTrace();
        }
        return emp;
    }
```

（8）创建 ProjectController 控制器，关键代码如下。

```
@Controller
public class ProjectController {
    @RequestMapping("/findProjectEmps")
    public ModelAndView findProjectEmps(int projectid) {
        ProjectDaoImpl projectDao=new ProjectDaoImpl();
        Project project=projectDao.findProjectEmps(projectid);
        ModelAndView mv=new ModelAndView();
        mv.addObject("project", project);
        mv.setViewName("forward:search9.jsp");
        return mv;
    }
}
```

（9）在 EmpController 控制器中添加如下方法。

```
@RequestMapping("/findEmpProjects")
public ModelAndView findEmpProjects(int empno) {
    EmpDaoImpl empDao=new EmpDaoImpl();
    Emp emp=empDao.findEmpProjects(empno);
    ModelAndView mv=new ModelAndView();
    mv.addObject("emp", emp);
    mv.setViewName("forward:search10.jsp");
    return mv;
}
```

（10）在 WebContent 下创建 search9.jsp 页面，用于查询项目和显示查询结果，关键代码如下。

```
<body>
<h1>查询项目信息</h1>
<form action="findProjectEmps" method="post">
项目编号：<input type="text" name="projectid" />
<input type="submit" value="查询" />
</form>
<c:if test="${not empty project }">
<hr/>
<h1>项目信息</h1>
<table border="1">
```

```
        <tr><th>项目编号</th><td>${project.projectid }</td></tr>
        <tr><th>项目名称</th><td>${project.projectname}</td>    </tr>
    </table>
     参与该项目的员工信息：<br/>
    <table border="1">
          <tr>
             <th>编号</th><th>姓名</th><th>职位</th><th>入职日期</th><th>经理
</th>
             <th>工资</th><th>奖金</th><th>部门</th><th>工作内容</th>
          </tr>
          <c:forEach items="${project.emps }" var="emp" varStatus="vs">
             <tr ${vs.count%2==1 ? "style='background-color:yellow'":
"style='background-color:white'" }>
                 <td>${emp.empno}</td><td>${emp.empname }</td><td>${emp.job}
 </td>
                 <td>${emp.hiredate.toLocaleString()}</td><td>${emp.manager}
</td>
                 <td>${emp.salary }</td>    <td>${emp.commission }</td>
                 <td>${emp.deptno }</td>    <td>${emp.content }</td>
             </tr>
          </c:forEach>
       </table>
    </c:if>
    </body>
```

（11）在 WebContent 下创建 search9.jsp 页面，用于查询员工和显示查询结果，关键代码如下。

```
    <body>
    <h1>查询员工信息</h1>
    <form action="findEmpProjects" method="post">
    员工编号：<input type="text" name="empno" />
    <input type="submit" value="查询" />
    </form>
    <c:if test="${not empty emp }">
    <hr/>
    <h1>员工信息</h1>
    <table border="1">
        <tr><th>编号</th><td>${emp.empno }</td></tr>
        <tr><th>姓名</th><td>${emp.empname}</td>    </tr>
        <tr><th>职位</th><td>${emp.job }</td>    </tr>
        <tr><th>入职日期</th><td>${emp.hiredate.toLocaleString()}</td>   </tr>
        <tr><th>经理</th><td>${emp.manager }</td>    </tr>
        <tr><th>工资</th><td>${emp.salary }</td>    </tr>
        <tr><th>奖金</th><td>${emp.commission }</td>    </tr>
        <tr><th>部门</th><td>${emp.deptno }</td>    </tr>
```

```
</table>
 该员工参与的项目信息：<br/>
<table border="1">
        <tr>
            <th>项目编号</th><th>项目名称</th>    <th>工作内容</th>
        </tr>
        <c:forEach items="${emp.projects}" var="project" varStatus="vs">
            <tr ${vs.count%2==1 ? "style='background-color:yellow'" :
                                        "style='background-color:white'" }>
                <td>${project.projectid}</td>
                <td>${project.projectname }</td>
                <td>${project.content }</td>
            </tr>
        </c:forEach>
    </table>
</c:if>
</body>
```

（12）运行测试，在浏览器地址栏中输入 http://localhost:8080/mybatis2/search9.jsp。输入项目编号 2，单击“查询”按钮，结果如图 8-20 所示。

图 8-20　某项目及参与员工信息查询结果

在浏览器地址栏中输入 http://localhost:8080/mybatis2/search10.jsp，再输入员工编号 1002，单击“查询”按钮，结果如图 8-21 所示。

图 8-21　某员工及参与项目信息查询结果

这样员工表 emp 与项目表 project 就实现了多对多查询。

视频讲解

8.7　分页查询

8.7.1　MyBatis 分页查询原理

分页查询需要确定查询第几页（pageNum），一页显示多少条记录（pageSize），这两项内容通常作为实际参数，传递给分页查询的相关方法。分页查询方法查询数据库后返回分页有关的各种信息，封装到一个分页信息类（如 PageInfo<T>）的实例对象中，再传递给前端展示。分页信息类通常包含如下一些属性：

- int pageNum：当前第几页（页码）。
- int pageSize：一页显示多少条记录。
- int size：记录总数。
- int startRow：起始显示的记录索引。
- int endRow：结束显示的记录索引。

- int pages：总页数。
- int prePage：上一页（页码）。
- int nextPage：下一页（页码）。
- boolean isFirstPage：是否第一页。
- boolean isLastPage：是否最后一页。
- boolean hasPreviousPage：是否有上一页。
- boolean hasNextPage：是否有下一页。
- List<T> list：当前页要展示的数据集合。

分页信息类的实例对象传递到前端后，可以将上述属性中的值取出来显示，从而达到分页的效果。

早期这种分页信息类（如 PageInfo<T>）需要自行设计，自行封装数据到分页信息类的对象，这个过程比较烦琐复杂。例如，已知页码 pageNumt 和一页显示的记录数 pageSize，首先计算表达式 (pageNum−1)*pageSize，将结果赋值给 startRow，然后将 startRow 和 pageSize 传递给 Select * from XXX limit startRow 这个 SQL 语句使用，从而使 pageSize 可以查询出该页对应的数据集合，并将其赋值给分页信息类 PageInfo 实例对象的 list 属性。记录总数则要通过 SQL 语句 select count(*) from XXX 进行查询，再赋值给分页信息类 PageInfo 实例对象的 list 属性。总页数则通过表达式 Math.ceil(size/pageSize) 进行计算，再将结果赋值给分页信息类 PageInfo 实例对象的 pages 属性。其他属性类似这样自行计算。

第三方的 PageHelper 类同样利用了上述原理快速实现分页，不再需要程序员自行计算各种分页信息，极大地简化了分页过程。

8.7.2 使用 PageHelper 实现分页

1. 项目中引入分页插件

（1）下载 pagehelper.jar 包

可以从下面的地址中下载最新版本的 JAR 包。

https://oss.sonatype.org/content/repositories/releases/com/github/pagehelper/pagehelper

http://repo1.maven.org/maven2/com/github/pagehelper/pagehelper/

（2）下载 jsqlparser.jar 包

由于使用了 sql 解析工具，还需要下载 jsqlparser.jar 包，注意需要和 PageHelper 依赖的版本一致。

http://repo1.maven.org/maven2/com/github/jsqlparser/jsqlparser/

2. 配置拦截器插件

在 MyBatis 配置 XML 中配置拦截器插件，代码如下所示。

```
<plugins>
    <plugin interceptor="com.github.pagehelper.PageInterceptor">
        <!-- 设置数据库类型 Oracle,Mysql,MariaDB,SQLite,Hsqldb,PostgreSQL
六种数据库-->
        <property name="helperDialect" value="mysql"/>
```

```
        </plugin>
    </plugins>
```

3. 在代码中使用分页

在查询有关的语句前面调用 PageHelper 的 startPage（int pageNum,int pageSize）方法，这样可以查询出分页后的结果。调用 PageHelper 的 startPage 方法后的第一个查询语句将会被分页，其后的语句不再受影响。

示例：

```
PageHelper.startPage(2, 3);
List<Emp> list = empDao.findAllEmps();
```

解释：第一行代码指使用 PageHelper 进行分页，查询第 2 页，每页显示 3 条记录。第二行代码是指查询所有员工信息，原本返回所有的员工数据，但由于被 PageHelper 拦截并分页，将只返回第 2 页的 3 条数据的集合。

最后将 list 作为 PageInfo 类的构造方法的参数，封装一个 PageInfo 实例对象 PageInfo<Emp> pageInfo=new PageInfo<Emp>(list)，再将其传递给前端。

8.7.3 分页实践

示例：将 Emp 中的数据分多页查询出来，每页显示 3 条。

实现步骤：

（1）导入 JAR 包 jsqlparser-3.1.jar 和 pagehelper-5.1.9.jar 到项目 mybatis2 中。

（2）在项目的 mybatis-config.xml 文件中配置拦截器插件，如图 8-22 所示。

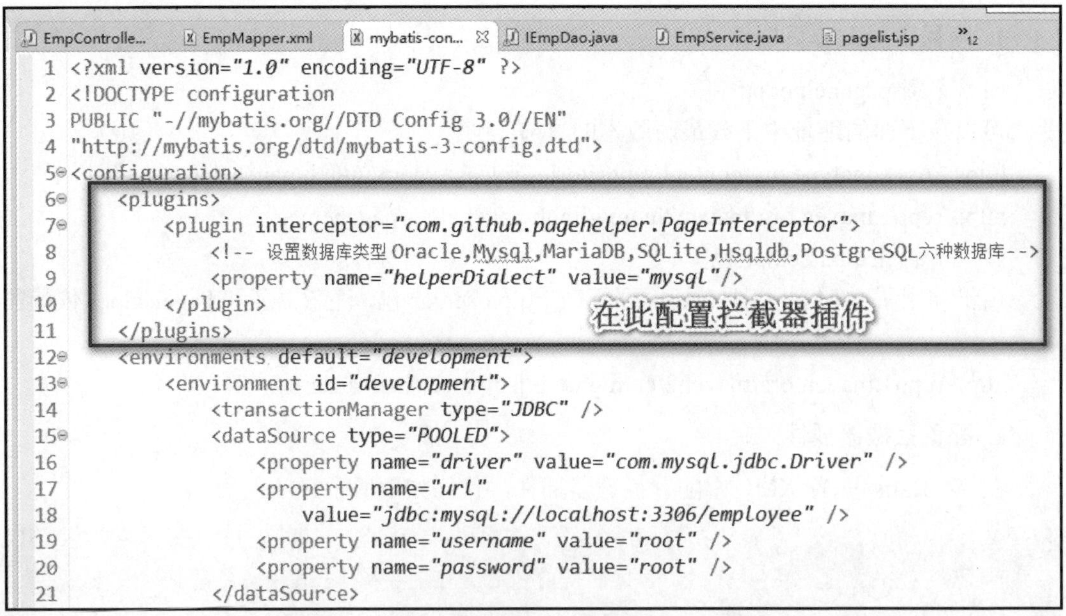

```
EmpControlle...  EmpMapper.xml  mybatis-con...  IEmpDao.java  EmpService.java  pagelist.jsp  »12
<?xml version="1.0" encoding="UTF-8" ?>
<!DOCTYPE configuration
PUBLIC "-//mybatis.org//DTD Config 3.0//EN"
"http://mybatis.org/dtd/mybatis-3-config.dtd">
<configuration>
    <plugins>
        <plugin interceptor="com.github.pagehelper.PageInterceptor">
            <!-- 设置数据库类型 Oracle,Mysql,MariaDB,SQLite,Hsqldb,PostgreSQL六种数据库-->
            <property name="helperDialect" value="mysql"/>
        </plugin>
    </plugins>                          在此配置拦截器插件
    <environments default="development">
        <environment id="development">
            <transactionManager type="JDBC" />
            <dataSource type="POOLED">
                <property name="driver" value="com.mysql.jdbc.Driver" />
                <property name="url"
                    value="jdbc:mysql://localhost:3306/employee" />
                <property name="username" value="root" />
                <property name="password" value="root" />
            </dataSource>
```

图 8-22　配置拦截器插件

（3）在 IEmpDao 接口中添加 findAllEmps 方法，在 EmpMapper.xml 添加以下代码。

```
<select id="findAllEmps" resultMap="empResultMap">
     select * from emp
</select>
```

（4）在 EmpDaoImpl 中重写 findAllEmps 方法，代码如下。

```
@Override
    public List<Emp> findAllEmps() {
        SqlSession session=null;
        List<Emp> list=new ArrayList<Emp>();
        try {
            session=MyBatisUtil.getSession();
            list=session.selectList("com.ssmbook2020.ch8.dao.IEmpDao.
findAllEmps");
            System.out.println(list.size());
        }catch (Exception e) {
            e.printStackTrace();
        }
        return list;
    }
```

（5）创建 com.seehope.service 包作为业务层，包下创建 EmpService 类，代码如下。

```
import com.github.pagehelper.PageHelper;
import com.github.pagehelper.PageInfo;
import com.ssmbook2020.ch8.dao.EmpDaoImpl;
import com.ssmbook2020.ch8.dao.IEmpDao;
import com.ssmbook2020.ch8.entity.Emp;

public class EmpService {
    public PageInfo<Emp> getAllEmpsPageInfo(int pageNum,int pageSize){
        IEmpDao empDao=new EmpDaoImpl();
        PageHelper.startPage(pageNum,pageSize);
        List<Emp> list=empDao.findAllEmps();
        PageInfo<Emp> pageInfo=new PageInfo<Emp>(list);
        return pageInfo;
    }
}
```

（6）在 EmpController 中创建 findAllEmps 方法，代码如下。

```
@RequestMapping("/empPage")
public ModelAndView findAllEmps(@RequestParam(value = "pageNum",
defaultValue = "1") int pageNum) {
    int pageSize=3;
    EmpService empService=new EmpService();
    PageInfo<Emp> pageInfo=empService.getAllEmpsPageInfo(pageNum, pageSize);
    ModelAndView mv=new ModelAndView();
    mv.addObject("pageInfo", pageInfo);
```

```
        mv.setViewName("forward:pagelist.jsp");
        return mv;
    }
```

（7）在 WebContent 下创建 pagelist.jsp 页面，关键代码如下。

```
    <body>
    <table border="1">
            <tr>
                <th>编号</th>
                <th>姓名</th>
                <th>职位</th>
                <th>入职日期</th>
                <th>经理</th>
                <th>工资</th>
                <th>奖金</th>
                <th>部门</th>
            </tr>
            <c:forEach items="${pageInfo.list }" var="emp" varStatus="vs">
                <tr ${vs.count%2==1 ? "style='background-color:yellow'" :
"style='background-color:white'" }>
                    <td>${emp.empno}</td>
                    <td>${emp.empname }</td>
                    <td>${emp.job }</td>
                    <td>${emp.hiredate.toLocaleString() }</td>
                    <td>${emp.manager }</td>
                    <td>${emp.salary }</td>
                    <td>${emp.commission }</td>
                    <td>${emp.deptno }</td>
                </tr>
            </c:forEach>
        </table>
        <div>
            <a href="empPage?pageNum=${pageInfo.isHasPreviousPage()?pageInfo.
getPrePage(): 1}">&lt;&lt;上一页</a>  
            第${pageInfo.getPageNum()}页/共${pageInfo.getPages()}页  
            <a href="empPage?pageNum=${pageInfo.isHasNextPage()?pageInfo.
getNextPage():pageInfo.getPages()}">下一页&gt;&gt;</a>  
        </div>
    </body>
```

（8）运行测试，在浏览器地址栏中输入 http://localhost:8080/mybatis2/empPage，结果如图 8-23 所示。

单击“下一页”，结果如图 8-24 所示。

再次单击“下一页”，结果如图 8-25 所示。至此，分页功能得到实现。

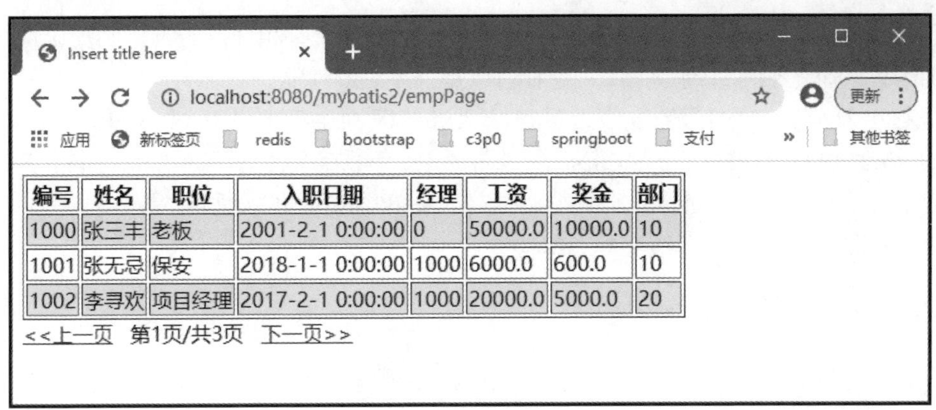

编号	姓名	职位	入职日期	经理	工资	奖金	部门
1000	张三丰	老板	2001-2-1 0:00:00	0	50000.0	10000.0	10
1001	张无忌	保安	2018-1-1 0:00:00	1000	6000.0	600.0	10
1002	李寻欢	项目经理	2017-2-1 0:00:00	1000	20000.0	5000.0	20

<<上一页 第1页/共3页 下一页>>

图 8-23　分页查询首页

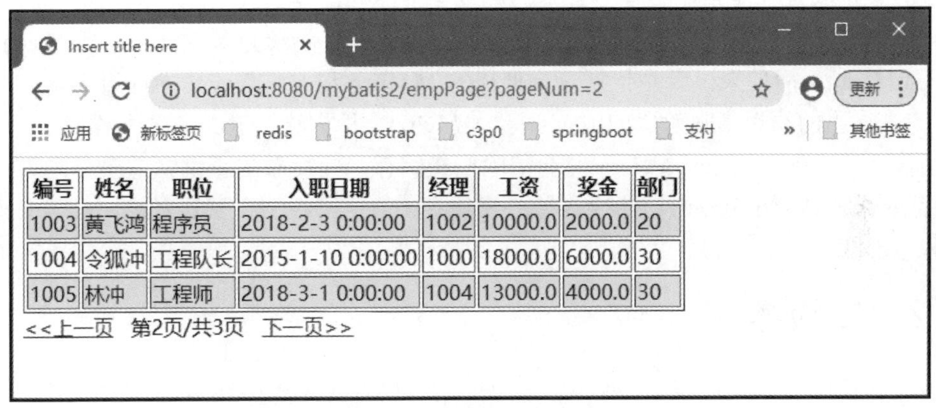

编号	姓名	职位	入职日期	经理	工资	奖金	部门
1003	黄飞鸿	程序员	2018-2-3 0:00:00	1002	10000.0	2000.0	20
1004	令狐冲	工程队长	2015-1-10 0:00:00	1000	18000.0	6000.0	30
1005	林冲	工程师	2018-3-1 0:00:00	1004	13000.0	4000.0	30

<<上一页 第2页/共3页 下一页>>

图 8-24　分页查询第 2 页

编号	姓名	职位	入职日期	经理	工资	奖金	部门
1006	鲁智深	工程师	2018-10-1 0:00:00	1004	11000.0	3000.0	30

<<上一页 第3页/共3页 下一页>>

图 8-25　分页查询第 3 页

8.8　缓　　存

视频讲解

MyBatis 每次查询数据库时，都会创建一个 SqlSession 对象并开启一个数据库会话。在一个数据库会话过程中，可能要多次查询相同的数据，获取相同的结果。如果每次都要连接数据库，执行 SQL 查询操作，会造成严重的资源浪费。为了解决这个问题，MyBatis

提供了缓存技术，将每次的查询结果存入缓存，如果下次有相同的查询，则直接从缓存获取查询结果，而无须再查询数据库，这样可以提升系统的性能。

缓存可以分为一级缓存和二级缓存，两者的区别主要是作用域不同。一级缓存是 SqlSession 级别的，在同一个 SqlSession 会话内读/写缓存有效，不同 SqlSession 无效。二级缓存的作用域为 sessionfactory，可以跨越不同 SqlSession 共享缓存。

下面分别深入学习一级缓存和二级缓存。

8.8.1 一级缓存

一级缓存默认开启，只要是同一个 SqlSession 的多次相同查询，第二次及以后的查询就会自动利用一级缓存，以提升系统性能。不同的 SqlSession 之间的缓存数据区域是互相不影响的。每次查询会先在缓存中查询，如果查询不到，再到数据库查询，然后把结果写到缓存中。MyBatis 的内部缓存使用一个 HashMap 来存储缓存数据，Key 为 hashcode+statementId+sql，简单来说就是依据查询使用的 Mapper 和方法名、传参等组成查询 Key，Value 为查询出的结果集映射成的 Java 对象。MyBatis 根据这个 Key 来查找缓存。

当执行 insert、update、delete 时会清除一级缓存，此后新的查询将重新查询数据库，当执行 commit 或者 rollback 时也会清除一级缓存。由于一级缓存默认开启，只需了解它的存在和原理即可。

8.8.2 二级缓存

二级缓存作用域为 SessionFactory，也称为 MapperStatement 级缓存。每生成一个命名空间 namespace（mapper.xml 映射文件）就会有一个缓存，不同的 SqlSession 之间的缓存是共享的。二级缓存默认关闭，需要手动开启，开启步骤如下。

（1）在 MyBatis 配置文件 mybatis-config.xml 中进行如下配置，开启二级缓存。

```
<settings>
    <setting name="cacheEnabled" value="true" />
</settings>
```

在 setting 里面开启二级缓存的作用域是全局性的，即所有查询都会启用二级缓存。如果只开启某一个 SQL 的二级缓存，可以在映射文件中相应的 select 标签里设置 useCache="true"。例如：

```
    <select id="findEmpById" resultType="com.ssmbook2020.ch8.entity.Emp"
parameterType="int" useCache="true">
            select * from emp where empno=#{empno}
    </select>
```

如果设置为 false，则这个 SQL 关闭二级缓存。

（2）在映射文件中添加<cache>配置，代码如下。

```
<cache eviction="LRU" flushInterval="100000" readOnly="true" size="1024"/>
```

其中，各个属性的含义说明如下。

① eviction：缓存回收策略，目前 MyBatis 提供以下策略。

- LRU：最近最少使用，移除最长时间不用的对象,为默认策略。
- FIFO：先进先出，按对象进入缓存的顺序进行移除。
- SOFT：软引用，移除基于垃圾回收器状态和软引用规则的对象。
- WEAK：弱引用，更积极地移除基于垃圾收集器状态和弱引用规则的对象。

② flushInterval：刷新间隔时间，单位为 ms，默认不设置，仅当执行 SQL 时刷新缓存。这里配置的是 100s 刷新。

③ size：引用数目，为正整数，代表缓存最多可以存储多少个对象，默认值为 1024。

④ readOnly：只读，意味着缓存数据只能读取而不能修改，默认值是 false。这样设置的好处是可以快速读取缓存，缺点是无法修改缓存。

为了避免出现脏数据，还要在每一次的 insert、update、delete 操作后都进行缓存刷新，也就是在 Statement 配置中配置 flushCache 属性，示例如下。

```
    <update id="updateEmp" parameterType="com.ssmbook2020.ch8.entity.Emp"
    flushCache="true">
           update emp set
           ename=#{empname},job=#{job},mgr=#{manager},hiredate=#{hiredate},
           sal=#{salary},comm=#{commission},deptno=#{deptno} where empno=
#{empno}
    </update>
```

（3）创建 POJO 并序列化

MyBatis 要求返回的 POJO 必须是可序列化的，也就是要求实现 Serializable 接口。

示例：Emp 类实现序列化接口。

```
    public class Emp implements Serializable
```

通过这些步骤就可以开启二级缓存，这样不同的 SqlSession 之间也能共享缓存数据，一定程度上提高了系统性能。虽然开启二级缓存能带来一定程度的好处，但实际业务中通常不建议开启。以下情况开启二级缓存可能会导致错误。

（1）针对一张表的某些操作不在独立的 namespace 下进行，导致两个命名空间下的数据不一致。

（2）多表查询操作使用缓存可能会出错。

习　题　8

1. 表与表之间主要有哪些关系？
2. MyBatis 实现一对多查询的关键步骤有哪些？
3. 什么是 MyBatis 一级缓存？
4. 什么是 MyBatis 二级缓存？

上机练习 2

在上机练习 1 创建的数据库 bookdb 中添加类别表 bookcategory，列有类别 id、类别名称 categoryname 和大类别 pid（可以为空，如《Java 程序设计》这本书的类别是计算机，大类别是工业技术）。修改 book 表，添加一个列 category 作为外键，即书的类别。添加若干测试数据。

1. 实现一对多查询，查询某个类别有哪些图书。
2. 实现多对一查询，查询一本书的同时查询该书的类别信息。
3. 实现自连接查询，查询图书类别及其大类别信息。

第9章 Spring事务管理

本章学习内容

- Spring 事务管理的概念；
- Spring 事务管理的核心接口；
- 声明式事务。

本章使用 Spring 来管理事务，学习事务的隔离级别、事务的提交与回滚等，掌握基于 XML 配置文件的事务管理方式和基于注解的事务管理方式。

9.1 事务管理的概念

所谓事务就是由一系列动作组成一个基本工作单元，这些动作要么全部完成，要么全部不完成，不允许只执行一部分而不执行另一部分的情况。例如，大家在生活中经常使用微信，微钱零钱来自绑定的银行卡，如果要往微信充值零钱 1000 元，就要从绑定的银行卡里扣掉 1000 元。这里包含了两个操作，一个是从银行卡里扣款，一个是往微信零钱里充值。如果这两个操作之间发生异常，就可能导致银行卡里的钱扣掉了，微信零钱却充值失败，显然用户“吃亏”了；反之，如果微信零钱充值成功，而银行卡扣款失败，则银行“吃亏”了。这些情况都是不允许发生的，即充值与扣款这两个操作应“打包”成一个不可分割的整体工作单元，要么全部成功，要么全部不成功，不允许只成功一个的情况。具体来说，如果先扣了款，但接下来发生异常，导致无法充值，这时应撤销先前的“扣款”操作，使数据重新回到一致状态，反之亦然。如果两个操作都成功了，这时才提交事务，永久改变数据。在 Java 开发中，事务管理用来保证数据的完整性与一致性。

事务管理有如下四个基本属性，简称 ACID。

- **原子性**（Atomicity）：事务是由一系列动作组成的原子操作。这一系列动作要么全部成功，要么全部失败，不允许部分成功部分失败。
- **一致性**（Consistency）：一旦事务完成，无论成功还是失败，所有数据必须处于一致的状态。如果事务执行过程中有部分操作完成后发生异常，应将已完成的操作撤销，以保证数据仍然处于一致状态。
- **隔离性**（Isolation）：多个事务在操作同一个数据时，彼此之间应互相隔离、互不影响。一个事务在查看数据时，所用数据必须是另一个事务修改前或修改后的数据，不能是中间状态的数据。
- **持久性**（Durability）：一旦事务完成，它对数据库的更改便是永久的，即使系统崩溃，数据库仍然可以恢复到最近一个事务成功结束时的状态。

视频讲解

9.2 Spring 事务管理的核心接口

Spring 通过三个核心接口来进行事务管理，分别是 TransactionDefinition 接口、TransactionStatus 接口和 PlatformTransactionManager 接口。TransactionDefinition 接口用于定

义事务属性与方法，PlatformTransactionManager 接口用于提供事务管理器。Spring 中的 spring-tx-4.3.4.RELEASE.jar 包是用于事务管理的依赖包，使用 Eclipse 工具展开该 JAR 包，再展开它的 org.springframework.transaction 包，可以看到上面所述的 3 个接口文件，如图 9-1 所示。

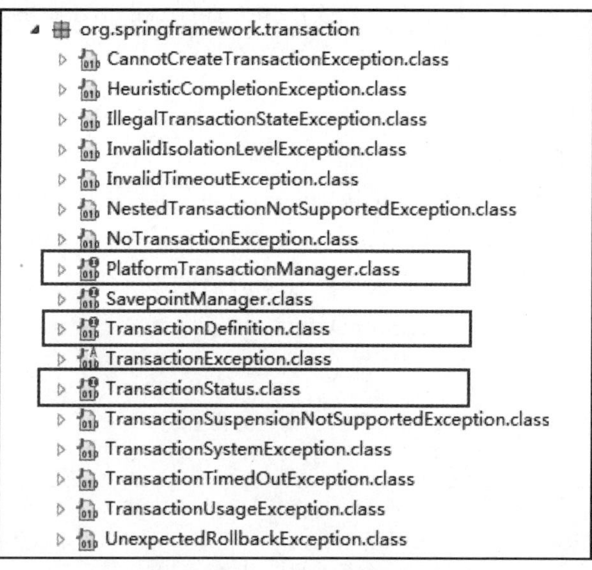

图 9-1　事务管理三大接口文件

9.2.1　TransactionDefinition 接口

TransactionDefinition 接口定义了事务的传播行为常量、事务的隔离级别常量和事务的超时时间常量。展开如图 9-1 所示的 TransactionDefinition.class 接口文件，可以看到该接口的所有属性与方法，如图 9-2 所示。接口 TransactionDefinition 提供了以下方法。

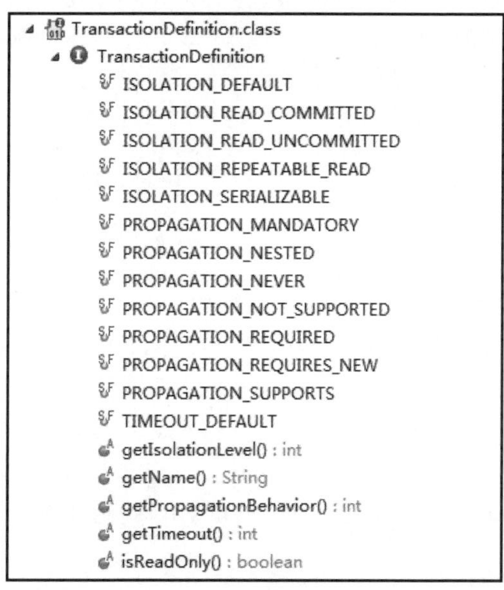

图 9-2　TransactionDefinition 接口详细信息

- int getIsolationLevel()：获得事务隔离级别；
- String getName()：获得事务名称；
- int getPropagationBehavior()：获得事务传播行为；
- int getTimeout()：获得事务超时时间，即必须在多少秒内完成；
- boolean isReadOnly()：事务是否只读，事务管理器根据该返回值进行优化，以确保事务是只读的。

1. 事务传播行为常量

在图 9-2 所示的 TransactionDefinition 接口中，以“PROPAGATION_”开头的常量属性定义了事务的传播行为（propagation behavior）。事务传播行为指的是当一个事务方法 B 调用另一个事务方法 A 时，应当明确规定事务如何传播，比如可以规定 A 方法继续在 B 方法的现有事务中运行，也可以规定 A 方法在新开启的事务中运行，将现有事务挂起并在新事务执行完毕后进行恢复。TransactionDefinition 接口定义了 7 种传播行为，各个行为说明如表 9-1 所示。

表 9-1　事务传播行为常量说明

传播行为名称	说　明
PROPAGATION_REQUIRED	表示受该传播行为约束的方法必须运行在事务中。如果当前已存在事务，方法将会在该事务中运行。如果当前没有事务，则会开启一个新的事务以执行该方法
PROPAGATION_SUPPORTS	表示受该传播行为约束的方法可以在没有事务的环境下运行。如果当前已存在事务，方法会支持该事务，并在该事务中运行
PROPAGATION_MANDATORY	表示受该传播行为约束的方法必须运行在事务中。如果当前不存在事务，则会抛出异常
PROPAGATION_REQUIRED_NEW	表示受该传播行为约束的方法必须启动一个新的事务来运行。如果当前已存在事务，则当前事务先挂起，直到新事务执行完毕
PROPAGATION_NOT_SUPPORTED	表示受该传播行为约束的方法不支持事务，不能运行在事务环境中。如果当前已存在事务，先将当前事务挂起，直到该方法执行完毕
PROPAGATION_NEVER	表示受该传播行为约束的方法不支持事务，不能运行在事务环境中。如果当前已存在事务，则会抛出异常
PROPAGATION_NESTED	表示如果当前已存在事务，那么受该传播行为约束的方法将会在嵌套事务中运行。内部嵌套的事务可以独立于当前事务进行单独地提交或回滚。如果当前事务不存在，那么其传播行为与 PROPAGATION_REQUIRED 一致

下面以PROPAGATION_REQUIRED和PROPAGATION_SUPPORTS这两个传播行为为例进行详细解释，其他传播行为以此类推。

1）PROPAGATION_REQUIRED

假定将方法 actionA()和方法 actionB()的事务传播行为都约束为 PROPAGATION_REQUIRED，然后在 main 方法中单独调用 actionB()，代码如下。

```
public static void main(String[] args){
 actionB();
```

```
}
```

这里并没有显式地定义事务的开始与提交，因此是一个非事务环境，当前不存在事务。在这种情况下，main 方法将会开启一个新的事务以执行 actionB()，相当于自行补充并执行以下代码。

```
public static void main(String[] args){
    Connection con=null;
    try{
        con = getConnection();
        //开启一个新的连接，即新的事务
        con.setAutoCommit(false);
        //方法调用
        actionB ();
        //提交事务
        con.commit();
    } Catch(RuntimeException ex) {
        //回滚事务
        con.rollback();
    } finally {
        //释放资源
        closeCon();
    }
}
```

如果方法 actionB()在运行中要调用方法 actionA()，则执行以下代码。

```
public static void main(String[] args){
    actionB(){
        actionA(){};
    }
}
```

这时由于方法 actionB()先执行，main 方法会自动开启一个事务运行，并将其视为当前事务。在运行过程中，方法 actionA()出现了，由于当前已存在事务，所以方法 actionA()就会加入到当前事务中运行（而不是再开启一个新的事务），相当于执行以下代码。

```
public static void main(String[] args){
    Connection con=null;
    try{
        con = getConnection();
        //开启一个新的连接，即新的事务
        con.setAutoCommit(false);
        //方法调用
        actionB(){
            actionA(){};
        }
        //提交事务
```

```
        con.commit();
    } Catch(RuntimeException ex) {
        //回滚事务
        con.rollback();
    } finally {
        //释放资源
        con.close();
    }
}
```

显然两个方法运行在同一个事务中。

2）PROPAGATION_SUPPORTS

假定将方法 actionA()的事务传播行为约束为 PROPAGATION_SUPPORTS，将方法 actionB()的事务传播行为约束为 PROPAGATION_REQUIRED，然后在 main 方法中单独调用 actionA()，代码如下。

```
public static void main(String[] args){
    actionA();
}
```

这里显然不存在事务，但 actionA()无须事务也能正常运行。

接着在 main 方法中调用 actionB()，方法 actionB()再调用 actionA()，代码如下。

```
public static void main(String[] args){
    actionB(){
         actionA(){};
    }
}
```

当前显然也不存在事务，但由于 actionB()的传播行为是 PROPAGATION_REQUIRED，系统会开启一个新的事务以执行 actionB()，并将其视为当前事务。然后执行过程中出现了 actionA()，这时它所在的环境显然是存在事务的，所以就加入到当前事务中。具体代码如下。

```
public static void main(String[] args){
    Connection con=null;
    try{
        con = getConnection();
        //开启一个新的连接，新的事务
        con.setAutoCommit(false);
        //方法调用
        actionB(){
            actionA(){};
        }
        //提交事务
        con.commit();
    } Catch(RuntimeException ex) {
        //回滚事务
```

```
        con.rollback();
    } finally {
        //释放资源
        con.close();
    }
}
```

2. 隔离级别常量

在图 9-2 所示的 TransactionDefinition 接口中，以“ISOLATION_”开头的属性定义了事务的隔离级别（isolation level），一共定义了 5 个事务隔离级别。隔离级别指的是一个事务可能受其他并发事务影响的程度。在应用程序中，多个事务并发运行且操作相同的数据，这可能会引起脏读、不可重复读和幻读等问题 。

- **脏读**（Dirty Read）：第一个事务访问并改写了数据，但还没有提交事务，这时第二个事务访问并读取了刚刚改写的数据。如果此时第一个事务回滚了，那第二个事务读取的数据就是无效的“脏数据”。
- **不可重复读**（Nonrepeatable Read）：第一个事务在其生命周期内多次查询同一个数据，在两次查询之间，第二个事务访问并改写了该数据，导致第一个事务两次查询同一个数据得到的结果不同。
- **幻读**（Phantom Read）：第一个事务在其生命周期内进行了两次查询条件相同的数据查询，第一次按该查询条件读取了几行数据，这时第二个事务访问并插入或删除了一些数据。然后第一个事务再次按同一条件查询时，发现多了一些原本不存在的记录或者原有的记录不见了。

不可重复读与幻读类似，即同样表现为两次读取的结果不同，但它们是有区别的。

不可重复读的主要问题是值被修改。经过同样的查询，再次读取出来后发现值不同了。例如，在事务 A 中，Alice 第一次读取了自己的成绩为 88，事务尚未提交，代码如下。

```
conA = getConnection();
select score from scores where studentname ="Alice";
```

在事务 B 中，老师修改 Alice 的成绩为 98，并提交了事务，代码如下。

```
conB = getConnection();
update scores set score = 98 where studentname ="Alice";
conB.commit();
```

在事务 A 中，Alice 再次查询自己的成绩时，成绩变为了 98（然后再提交事务），代码如下。

```
select score from scores where studentname ="Alice";
conA.commit();
```

这样在一个事务中前后两次读取的结果不一致，导致了“不可重复读”。

幻读的主要问题在于新增或者删除记录。查询条件相同，但同一个事务中第一次和第二次读出来的记录数不同。

例如，目前成绩高于 90 分的学生有 10 人，事务 A 读取所有成绩高于 90 分的学生。

```
conA = getConnection();
select * from scores where score>90;
```

该查询共输出 10 条记录，这时另一个事务向 scores 表插入了一条新的成绩记录，成绩高于 90 分。

```
conB = getConnection();
Insert into scores(studentname,scores) values("LiSi",96);
conB.commit();
```

接着事务 A 再次读取成绩高于 90 分的学生（然后再提交事务）。

```
Select * from scores where score>90;
conA.commit();
```

此时一共输出了 11 条记录，显然比上次多了一条数据，这样就产生了“幻读”。

为了解决并发事务问题，Spring 定义了如表 9-2 所示的 5 种隔离级别。

表 9-2　隔离级别常量

隔离级别名称	含　义
ISOLATION_DEFAULT	默认的隔离级别，使用数据库默认的事务隔离级别。其余 4 个则对应 JDBC 的隔离级别
ISOLATION_READ_UNCOMMITTED	读未提交，最低的隔离级别。允许另一个事务读取该事务未提交的数据，可能会导致脏读、幻读或不可重复读
ISOLATION_READ_COMMITTED	读已提交。一个事务只能读取另一个事务已经提交的数据，可以有效防止脏读，但不能防止不可重复读和幻读
ISOLATION_REPEATABLE_READ	可重复读。可以防止脏读和不可重复读，但不能防止幻读
ISOLATION_SERIALIZABLE	可序列化，最高的隔离级别。事务按顺序执行，不存在并发问题，可以防止脏读、幻读或不可重复读。但该隔离级别效率最低，并不提倡使用

3. 事务默认超时时间常量

事务默认超时时间常量对应图 9-2 中的 TIMEOUT_DEFAULT 常量。为了使应用程序运行良好，事务不能运行太长的时间。由于事务可能涉及对数据库的锁定，长时间的事务会占用数据库资源，降低系统的性能，所以需要限定事务的超时时间。事务超时相当于事务的一个定时器，如果事务没有在特定时间内执行完毕，那么就会自动进行回滚，而不是一直等待事务结束。

4. 回滚规则

Spring 定义了会导致事务回滚或不回滚的异常。默认情况下，事务只有在遇到运行异常时才会回滚，在遇到检查型异常时则不会回滚。可以声明事务在遇到特定的检查型异常时进行回滚，反之，也可以声明事务遇到特定的运行异常时不回滚。

9.2.2　TransactionStatus 接口

TransactionStatus 接口描述的是一些控制事务执行和查询事务状态的方法，在事务回滚或提交时需要应用对应的事务状态。展开如图 9-1 所示的 TransactionStatus.class 接口文件，可以看到如图 9-3 所示的接口方法。

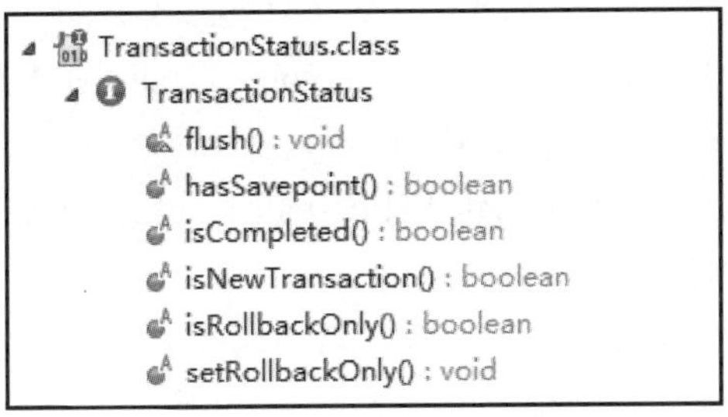

```
TransactionStatus.class
    TransactionStatus
        flush() : void
        hasSavepoint() : boolean
        isCompleted() : boolean
        isNewTransaction() : boolean
        isRollbackOnly() : boolean
        setRollbackOnly() : void
```

图 9-3　TransactionStatus 接口结构

对图 9-3 所示的接口方法解释如下。

```
public interface TransactionStatus{
    boolean isNewTransaction(); // 是否是新的事务
    boolean hasSavepoint();     // 是否有恢复点
    void setRollbackOnly();     // 设置为只回滚
    boolean isRollbackOnly();   // 是否为只回滚
    boolean isCompleted;        // 是否已完成
}
```

9.2.3 PlatformTransactionManager 接口

Spring 事务管理器的接口是 org.springframework.transaction.PlatformTransactionManager，Spring 通过该接口为 JDBC、MyBatis、Hibernate 等平台都提供了对应的事务管理器。此接口的结构如图 9-4 所示。

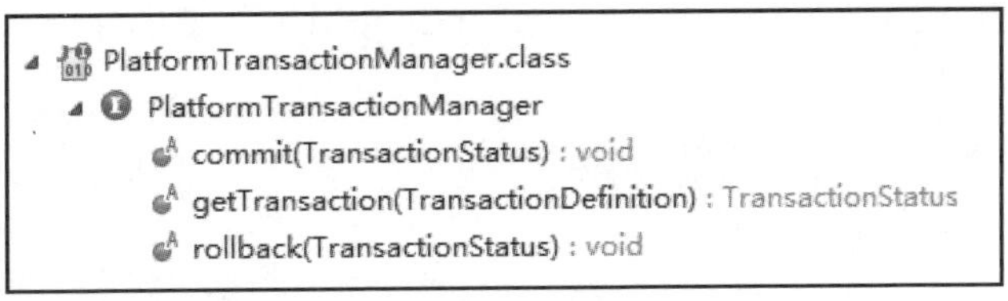

```
PlatformTransactionManager.class
    PlatformTransactionManager
        commit(TransactionStatus) : void
        getTransaction(TransactionDefinition) : TransactionStatus
        rollback(TransactionStatus) : void
```

图 9-4　PlatformTransactionManager 接口结构

该接口的各个方法解释如下。

```
public interface PlatformTransactionManager()...{
    // 传入 TransactionDefinition 参数获取 TransactionStatus 对象，用于获取事务状
态信息
    TransactionStatus getTransaction(TransactionDefinition definition)
throws TransactionException;
    // 提交事务
    void commit(TransactionStatus status) throws TransactionException;
    // 回滚事务
    void rollback(TransactionStatus status) throws TransactionException;
}
```

Spring 并不直接管理事务，而是提供了多种事务管理器，通过 JDBC、MyBatis、Hibernate 或者 JTA 等持久化机制所提供的相关平台框架的事务来实现。

PlatformTransactionManager 接口常用以下 3 个实现类。

- org.springframework.jdbc.datasource.DataSourceTransactionManager 实现类：使用 JDBC 或 MyBatis 进行数据持久化时使用。
- org.springframework.orm.hibernate4.HibernateTransactionManager 实现类：使用 Hibernate 进行数据持久化时使用。
- org.springframework.transaction.jta.JtaTransactionManager 实现类：用于配置全局事务管理器。

1. JDBC 事务

如果应用程序直接使用 JDBC，或者使用 MyBatis 来进行持久化，DataSourceTransactionManager 实现类会为用户处理事务边界。这时需要在 Spring 配置文件 applicationContext.xml 中进行以下配置，以将其装配到应用程序的上下文定义中。

```
    <bean id="transactionManager" class="org.springframework.jdbc.datasource.
DataSourceTransactionManager">
           <property name="dataSource" ref="dataSource" />
    </bean>
```

DataSourceTransactionManager 实际上是通过调用 java.sql.Connection 连接对象来管理事务，该对象通过 DataSource 进行获取。通过调用连接的 commit()方法来提交事务，通过调用连接的 rollback()方法进行回滚。

2. Hibernate 事务

如果应用程序的持久化是通过 Hibernate 实现的，就使用 HibernateTransactionManager 实现类。此时需要在 Spring 配置文件 applicationContext.xml 中进行以下配置，以将其装配到应用程序的上下文定义中。

```
    <bean id="transactionManager" class="org.springframework.orm.hibernate3.
HibernateTransactionManager">
           <property name="sessionFactory" ref="sessionFactory" />
    </bean>
```

HibernateTransactionManager 将事务管理的责任委托给 org.hibernate.Transaction 对象。当事务成功完成时，调用 Transaction 对象的 commit()方法提交事务；若失败，则调用 rollback()方法回滚事务。

3. Java 原生 API 事务

如果应用程序跨越了两个或多个不同的数据源，就需要使用 JtaTransactionManager 实现类。这时需要在 Spring 配置文件 applicationContext.xml 中进行以下配置，以将其装配到应用程序的上下文定义中。

```
    <bean id="transactionManager" class="org.springframework.transaction.jta.
```

```
JtaTransactionManager">
        <property name="transactionManagerName"
 value="java:/TransactionManager" />
     </bean>
```

JtaTransactionManager 将事务管理的责任委托给 javax.transaction.UserTransaction 和 javax.transaction.TransactionManager 两个对象，两个对象各有分工。当事务成功完成就调用 UserTransaction.commit()方法提交事务，若失败就调用 UserTransaction.rollback()方法回滚事务。

9.3 声明式事务

9.3.1 编程式和声明式事务的区别

视频讲解

Spring 支持编程式事务和声明式事务。编程式事务直接在主业务代码中精确定义事务的边界，声明式事务则基于 AOP 方式，将主业务操作与事务规则进行解耦。

编程式事务以硬编码的方式嵌入主业务代码里面，好处是能提供更加详细的事务管理，而声明式事务则基于 AOP 方式，能在不影响业务代码的具体实现情况下实现事务管理。但由于编程式事务的主业务与事务代码混在一起，不易分离且耦合度高，不利于维护与重用，所以比较常用的是声明式事务。

声明式事务有两种具体的实现方式：基于 XML 配置文件的方式和基于注解的方式。

9.3.2 基于 XML 配置文件的事务管理

视频讲解

基于 XML 配置文件的事务管理的 applicationContext.xml 配置文件模板如下。

```
<?xml version="1.0" encoding="UTF-8"?>
<beans xmlns="http://www.springframework.org/schema/beans"
   xmlns:xsi="http://www.w3.org/2001/XMLSchema-instance"
   xmlns:aop="http://www.springframework.org/schema/aop"
   xmlns:tx="http://www.springframework.org/schema/tx"
   xmlns:context="http://www.springframework.org/schema/context"
   xsi:schemaLocation="
      http://www.springframework.org/schema/beans
      http://www.springframework.org/schema/beans/spring-beans.xsd
      http://www.springframework.org/schema/context
      http://www.springframework.org/schema/context/spring-context.xsd
      http://www.springframework.org/schema/tx
      http://www.springframework.org/schema/tx/spring-tx.xsd
      http://www.springframework.org/schema/aop
      http://www.springframework.org/schema/aop/spring-aop.xsd">
   <!-- 配置数据源 -->
   <bean id="dataSource" class="org.springframework.jdbc.datasource.
```

```
DriverManagerDataSource">
            <property name="driverClassName">
                <value>com.mysql.jdbc.Driver</value>
            </property>
            <property name="url">
                <value>jdbc:mysql://localhost:3306/weixin</value>
            </property>
            <property name="username">
                <value>root</value>
            </property>
            <property name="password">
                <value>root</value>
            </property>
        </bean>
        <!-- 定义事务管理器 -->
        <bean id="txManager"
            class="org.springframework.jdbc.datasource.
DataSourceTransactionManager">
            <property name="dataSource" ref="dataSource" />
        </bean>
        <!-- 编写事务通知 -->
        <tx:advice id="txAdvice" transaction-manager="txManager">
            <tx:attributes>
                <tx:method name="save*" propagation="REQUIRED" />
                <tx:method name="add*" propagation="REQUIRED" />
                <tx:method name="insert*" propagation="REQUIRED" />
                <tx:method name="delete*" propagation="REQUIRED" />
                <tx:method name="update*" propagation="REQUIRED" />
                <tx:method name="search*" propagation="SUPPORTS"  read-only=
"true"/>
                <tx:method name="select*" propagation="SUPPORTS"  read-only=
"true"/>
                <tx:method name="find*" propagation="SUPPORTS"  read-only=
"true"/>
                <tx:method name="get*" propagation="SUPPORTS"  read-only=
"true"/>
            </tx:attributes>
        </tx:advice>
        <!-- 编写AOP,让Spring自动将事务切入到目标切点 -->
        <aop:config>
            <!-- 定义切入点 -->
            <aop:pointcut id="txPointcut"
                expression="execution(* com.service.*.*(..))" />
            <!-- 将事务通知与切入点组合 -->
            <aop:advisor advice-ref="txAdvice" pointcut-ref="txPointcut" />
        </aop:config>
```

```
<!-- 省略各种 Bean 的创建 -->
</beans>
```

代码解释：使用 AOP 方法将 com.service 包下的所有类和所有方法都纳入事务管理。其中，save 开头的方法，add 开头的方法，以及 insert、update、delete 开头的方法，事务传播行为设置为 REQUIRED，即必须运行于事务中。其他开头的方法，事务传播行为设置为 SUPPORTS，即有无事务环境均可运行，并且设置为只读。一般情况下对数据库进行增、删、改的方法要设置为 REQUIRED，查询数据库的方法设置为 SUPPORTS 及只读。

示例：模拟微信中的零钱充值业务,张三使用微信绑定的银行卡充值 1000 元。这其实包含了两个动作，一是从银行卡扣款 1000 元，二是往微信中添加 1000 元。这两个动作应同时完成或同时不完成，否则数据会不一致，但如果这两个动作之间发生异常呢？如果有问题，该如何解决？(项目源码参见:第 9 章/使用 XML 实现事务管理/weixin)

实现步骤：

（1）在 MySQL 数据库中创建数据库 weixin，创建数据表 weixinmoney 用于存储微信零钱，创建 card 用于存储银行卡账号余额。

初始情况下 weixinmoney 表的数据如图 9-5 所示，即微信零钱初始为 0。

card 表的初始数据如图 9-6 所示，即卡里有 10000 元。

	accountname	amount
□	张三	0
*	(NULL)	(NULL)

图 9-5　weixinmoney 表初始数据

	cardname	cardamount
□	张三	10000
*	(NULL)	(NULL)

图 9-6　card 表初始数据

（2）在 Eclipse 中新建 dynamic web project 项目，项目名称为 weixin，导入 Spring 包，连接 MySQL 数据库的 JAR 包。在 src 下新建 applicationContext.xml 配置文件，内容与上述模板代码基本一致，但需要先注释掉事务管理有关代码，暂时不做事务管理，并补充以下代码，创建 DAO 层和业务层的 Bean。

```
<!-- 配置 JdbcTemplate 模板 -->
<bean id="jdbcTemplate" class="org.springframework.jdbc.core.JdbcTemplate">
    <property name="dataSource" ref="dataSource" />
</bean>
<!-- 配置 DAO 层,注入 jdbcTemplate 属性值 -->
<bean id="weixinmoneyDao" class="com.dao.WeixinmoneyDaoImpl">
    <property name="jdbcTemplate" ref="jdbcTemplate"/>
</bean>
<bean id="cardDao" class="com.dao.CardDaoImpl">
    <property name="jdbcTemplate" ref="jdbcTemplate"/>
</bean>
<!-- 配置 SERVICE 层,注入 accountDao 属性值 -->
<bean id="weixinmoneyService" class="com.service.WeixinmoneyServiceImpl">
    <property name="weixinmoneyDao " ref="weixinmoneyDao "/>
</bean>
```

（3）在 com.dao 包下创建接口 WeixinmoneyDao，代码如下。

```
public interface WeixinmoneyDao{
    public void updateWeixin(String accountName,int amount);
}
```

（4）在 com.dao 包下创建接口 WeixinmoneyDao 的实现类 WeixinmoneyDaoImpl，代码如下。

```
public class WeixinmoneyDaoImpl implements WeixinmoneyDao {
    JdbcTemplate jdbcTemplate;
    public JdbcTemplate getJdbcTemplate() {
        return jdbcTemplate;
    }
    public void setJdbcTemplate(JdbcTemplate jdbcTemplate) {
        this.jdbcTemplate = jdbcTemplate;
    }
    @Override
    public void updateWeixin(String accountName, int amount) {
        jdbcTemplate.update("update weixinmoney set amount=amount+? where
accountname=?",amount,accountName);
    }
}
```

（5）在 com.dao 包下创建接口 CardDao，代码如下。

```
public interface CardDao {
    public void updateCard(String cardName,int cardAmount);
}
```

（6）在 com.dao 包下创建接口 CardDao 的实现类 CardDaoImpl，代码如下。

```
public class CardDaoImpl implements CardDao {
    JdbcTemplate jdbcTemplate;
    public JdbcTemplate getJdbcTemplate() {
        return jdbcTemplate;
    }
    public void setJdbcTemplate(JdbcTemplate jdbcTemplate) {
        this.jdbcTemplate = jdbcTemplate;
    }
    @Override
    public void updateCard(String cardName,int cardAmount) {
        jdbcTemplate.update("update card set cardamount=cardamount-? where
cardname=?",cardAmount,cardName);
    }
}
```

（7）在 com.service 包下创建接口 WeixinmoneyService，代码如下。

```
public interface WeixinmoneyService {
```

```
    public void chongZhi(String accountName,int amount);
}
```

（8）在 com.service 包下创建接口 WeixinmoneyService 的实现类 WeixinmoneyServiceImpl 如下，在 chongZhi()方法的充值过程中模拟发生异常。

```
public class WeixinmoneyServiceImpl implements WeixinmoneyService{
    private CardDao cardDao;
    private WeixinmoneyDao weixinmoneyDao;
    public CardDao getCardDao() {
        return cardDao;
    }
    public void setCardDao(CardDao cardDao) {
        this.cardDao = cardDao;
    }
    public WeixinmoneyDao getWeixinmoneyDao() {
        return weixinmoneyDao;
    }
    public void setWeixinmoneyDao(WeixinmoneyDao weixinmoneyDao) {
        this.weixinmoneyDao = weixinmoneyDao;
    }

//模拟充值业务
    @Override
    public void chongZhi(String accountName, int amount) {
        //先从卡里扣款
        cardDao.updateCard(accountName, amount);
        //模拟发生异常
        int num=1/0;
        //再把款项充入微信
        weixinmoneyDao.updateWeixin(accountName, amount);
    }
}
```

（9）在 com.test 包下创建测试类 TestWeixin，代码如下。

```
public class TestWeixin {
    public static void main(String[] args) {
        ApplicationContext context=new ClassPathXmlApplicationContext
("applicationContext.xml");
        WeixinmoneyService weixinmoneyService=(WeixinmoneyService) context.
getBean("weixinmoneyService");
        System.out.println("-------模拟张三的微信充值1000元-------");
        weixinmoneyService.chongZhi("张三", 1000);
    }
}
```

查看数据库发现 card 里的钱被扣掉了 1000 元，但 weixingmoney 却没充到钱，如图 9-7

和图 9-8 所示。很明显这是不合理的，造成了数据的不一致。即使在发生异常的情况下，程序也应该保持数据的一致性，这就需要用到事务管理，下面将对该项目进行改造。

	cardname	cardamount
□	张三	9000
*	(NULL)	(NULL)

图 9-7　执行后的 card 表数据

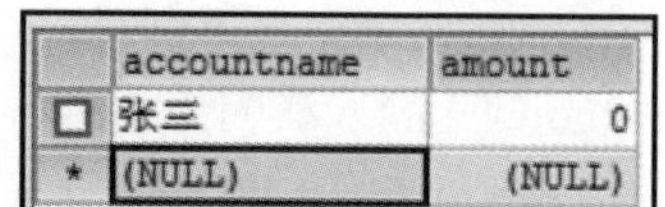

	accountname	amount
□	张三	0
*	(NULL)	(NULL)

图 9-8　执行后的 weixinmoney 表数据

（10）使用 XML 配置方式进行事务管理，在 Spring 配置文件夹下的 applicationContext .xml 文件中添加事务管理有关代码。具体内容见步骤（2），取消注释掉的内容即可。

再次运行测试，查看数据库，发现两个表都没变化，证明事务管理成功。其内在原理是，首先从卡里扣款，这个操作虽然成功了，但接下来发生异常，这时事务回滚，把先前的操作撤销，从而保证数据仍然处于一致状态。

视频讲解

9.3.3　注解式事务管理

使用@Transactional 在方法上添加注解可以对该方法进行事务管理。当然也可以将@Transactional 注解添加到类上，这样就表示对整个类的所有方法都进行事务管理。使用注解方式进行事务管理，XML 配置文件内容会简化很多，看起来不会那么“臃肿”。

要使用@Transactional 注解，首先要在 applicationContext.xml 配置文件中注册事务注解驱动，代码如下所示。

```
<tx:annotation-driven transaction-manager="txManager"/>
```

@Transactional 注解有多个参数，其含义与作用如表 9-3 所示。

表 9-3　@Transactional 注解参数说明

参数名	含义与作用
propagation	事务传播行为，枚举类型，取值参考表 9-1，默认为 Propagation .REQUIRED
isolation	事务隔离级别，枚举类型，取值参考表 9-2，默认为 Isolation.DEFAULT
readOnly	事务是否只读，可设置为 true 或 false
value	指定事务管理器，默认为""，即 XML 配置文件中注册的事务管理器
rollbackFor	指定异常类，遇到该非运行异常类时将进行回滚事务
rollbackForClassName	指定异常类，遇到这些异常类时都将进行回滚事务，指定字符串中的类名
noRollbackFor	指定异常类，遇到该运行异常类时将强制不回滚事务
noRollbackForClassName	指定异常类，遇到这些异常类时都将不回滚事务，指定字符串中的类名
timeout	指定强制回滚前事务可以占用的时间，默认为 TransactionDefiantion. TIMEOUT_DEFAULT

下面将 9.3.2 节的项目 weixin 改造为用注解方式实现事务管理。

（1）复制项目 weixin 为 weixin2，修改 applicationCongtext.xml 配置文件如下。

```
<?xml version="1.0" encoding="UTF-8"?>
<beans xmlns="http://www.springframework.org/schema/beans"
```

```
    xmlns:xsi="http://www.w3.org/2001/XMLSchema-instance"
    xmlns:aop="http://www.springframework.org/schema/aop"
    xmlns:tx="http://www.springframework.org/schema/tx"
    xmlns:context="http://www.springframework.org/schema/context"
    xsi:schemaLocation="
        http://www.springframework.org/schema/beans
        http://www.springframework.org/schema/beans/spring-beans.xsd
        http://www.springframework.org/schema/context
        http://www.springframework.org/schema/context/spring-context.xsd
        http://www.springframework.org/schema/tx
        http://www.springframework.org/schema/tx/spring-tx.xsd
        http://www.springframework.org/schema/aop
        http://www.springframework.org/schema/aop/spring-aop.xsd">
    <!-- 配置数据源 -->
    <bean id="dataSource" class="org.springframework.jdbc.datasource.
DriverManagerDataSource">
        <property name="driverClassName">
            <value>com.mysql.jdbc.Driver</value>
        </property>
        <property name="url">
            <value>jdbc:mysql://localhost:3306/weixin?characterEncoding=
utf8</value>
        </property>
        <property name="username">
            <value>root</value>
        </property>
        <property name="password">
            <value>root</value>
        </property>
    </bean>

    <!-- 配置 JdbcTemplate 模板 -->
    <bean id="jdbcTemplate" class="org.springframework.jdbc.core.
JdbcTemplate">
        <property name="dataSource" ref="dataSource" />
    </bean>

    <!-- 配置 DAO 层,注入 jdbcTemplate 属性值 -->
    <bean id="cardDao" class="com.dao.CardDaoImpl">
        <property name="jdbcTemplate" ref="jdbcTemplate"/>
    </bean>

    <!-- 配置 DAO 层,注入 jdbcTemplate 属性值 -->
    <bean id="weixinmoneyDao" class="com.dao.WeixinmoneyDaoImpl">
        <property name="jdbcTemplate" ref="jdbcTemplate"/>
```

```
    </bean>

    <!-- 配置 Service 层,注入 carDao 及 weixinmoneyDao 属性值 -->
    <bean id="weixinmoneyService" class="com.service.WeixinmoneyServiceImpl">
        <property name="cardDao" ref="cardDao"/>
        <property name="weixinmoneyDao" ref="weixinmoneyDao"/>
    </bean>

    <!-- 定义事务管理器 -->
    <bean id="txManager"
        class="org.springframework.jdbc.datasource.
DataSourceTransactionManager">
        <property name="dataSource" ref="dataSource" />
    </bean>
    <tx:annotation-driven transaction-manager="txManager"/>
</beans>
```

（2）在 WeixinmonyServiceImpl 类的 chongZhi 方法上添加注解，代码如下。

```
@Override
@Transactional(propagation=Propagation.REQUIRED,
isolation=Isolation.DEFAULT,readOnly=false)
public void chongZhi(String accountName, int amount) {
    //先从卡里扣款
    cardDao.updateCard(accountName, amount);
    //模拟发生异常
    int num=1/0;
    //再把款项充入微信
    weixinmoneyDao.updateWeixin(accountName, amount);
}
```

这里注解的意思是将方法 chongZhi()纳入事务管理，传播行为设置为 REQUIRED，即必须运行在事务环境中，隔离级别为默认，非只读。

运行测试，发现结果不变，证明注解方式事务管理成功。

习　题　9

1．事务的基本属性有哪些？
2．并发事务会导致哪些问题？
3．简述事务的 7 个传播行为。
4．简述事务的 5 个隔离级别。
5．PlatformTransactionManager 接口的主要实现类有哪些？

上机练习 3

创建数据表 account，设置账户名 accountname 和余额 balance 两个列，初始值如图 9-9 所示。

	accountname	balance
□	张无忌	10000
□	李寻欢	10000
*	(NULL)	(NULL)

图 9-9　账户初始值

创建项目，使用 Spring JDBC 操作数据库，模拟张无忌向李寻欢转账 3000 元，但业务中间发生异常。先不用事务管理，运行测试结果，观察数据库的变化。该过程有无问题？如有问题，应如何解决（使用 XML 配置方式）？

第10章 前端框架Vue基础

本章学习内容

- Vue 简介；
- IntelliJ IDEA 开发环境；
- Vue 的常用指令；
- Vue 的 AJAX 异步操作。

10.1 Vue 简 介

1. Vue 的基本概念

Vue 是一套构建用户界面的渐进式前端框架。什么是渐进式框架呢？ 与之对应的是全家桶式框架，即一个框架具备所有功能，无论是否用到，开发时都需要加载框架的全部功能，这样会导致过多地占用系统资源。渐进式框架可以根据需要导入功能，从而按需占用资源。

Vue 的核心库只关注视图层，不仅易于上手，而且便于与第三方库或既有的项目进行整合。Vue 的官网首页如图 10-1 所示，其网址为 https://cn.vuejs.org/，上面有中文教程。

图 10-1　Vue 的官网

2. Vue 的组成和特点

Vue 的核心库包含以下两个部分。

（1）视图 View：负责 HTML 页面的渲染，主要由 HTML 和 CSS 构成。

（2）模型 Model：负责业务数据的存储及数据的处理。

Vue 的特点如下。

（1）易用：有 HTML、CSS 和 JavaScript 基础的读者可以快速上手。

（2）灵活：简单小巧的核心，采用渐进式的技术栈，按需导入，足以应对任何规模的应用。

（3）性能优异：占用很小的运行内存空间，在浏览器中快速虚拟 DOM 元素树，最省心的系统优化。

正是因为 Vue 的出现，让功能强大、驰骋江湖多年的 jQuery 面临淡出历史舞台的命运。需要注意的是，Vue 不支持 IE8 及以下的浏览器版本，因为 Vue 使用了 IE8 无法模拟的 ECMAScript 5 特性。Vue 支持所有兼容 ECMAScript 5 及以上的浏览器，如 Chrome、

FireFox、Edge、Safari 和 Opera 等。

10.2 IntelliJ IDEA 开发环境

1. 开发工具

本书前面的章节中所使用的集成开发环境（IDE）是 MyEclipse，最后两章将使用 IntelliJ IDEA（以下简称 IDEA），这是目前不少企业使用的另一个主流开发工具。两个开发工具希望读者都能有所涉及。IDEA 功能上比 MyEclipse 要强大不少，但占用的系统资源也更多。

IntelliJ IDEA 是一个商用软件，官网首页如图 10-2 所示，官网提供了免费的试用版，有 30 天的试用期。读者可以在 https://www.jetbrains.com/IDEA/自行下载。本书使用的是 2019.1 的版本。

图 10-2　IntelliJ IDEA 下载

2. IDEA 创建工程

（1）如图 10-3 所示为 IDEA 的启动界面，选择 Create New Project 创建一个工程，也可以选择第二项 Import Project，从 Eclipse 等开发工具中导入以前的工程，它会转换成 IntelliJ IDEA 工程打开。

图 10-3　创建工程

（2）要将多个案例放在一个工程中，所以先创建一个空的工程，每个案例作为工程中的一个模块，如图 10-4 所示。

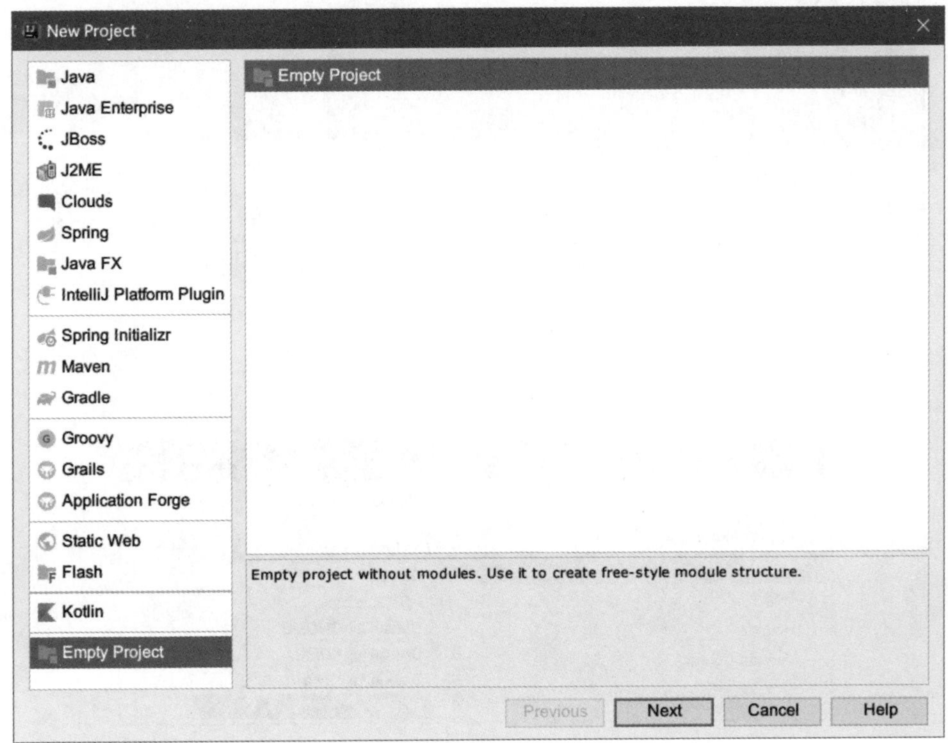

图 10-4　创建空工程

（3）为工程命名并设置它的存储位置，如图 10-5 所示。

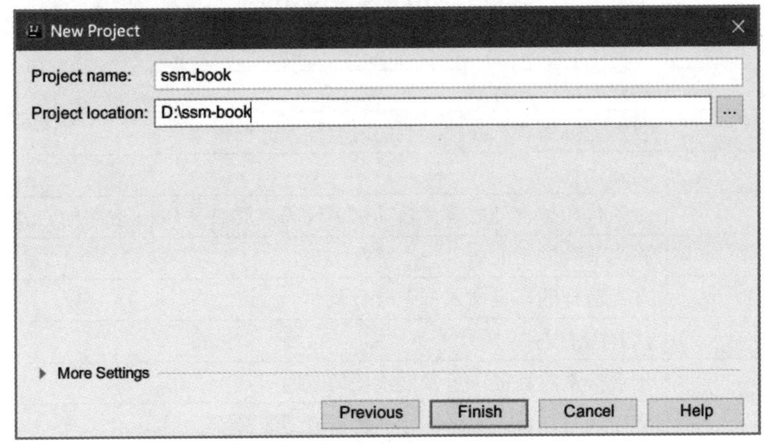

图 10-5　为工程命名

（4）在 Modules 窗口中选择取消，先不创建任何模块，如果 10-6 所示。

（5）如果习惯 Eclipse 的操作快捷键，可以将所有的快捷键设置成 Eclipse 操作习惯。如图 10-7 所示，选择 File 菜单→Settings 选项→Keymap 选项→Eclipse 选项，不过还是建议先熟悉新的开发工具。

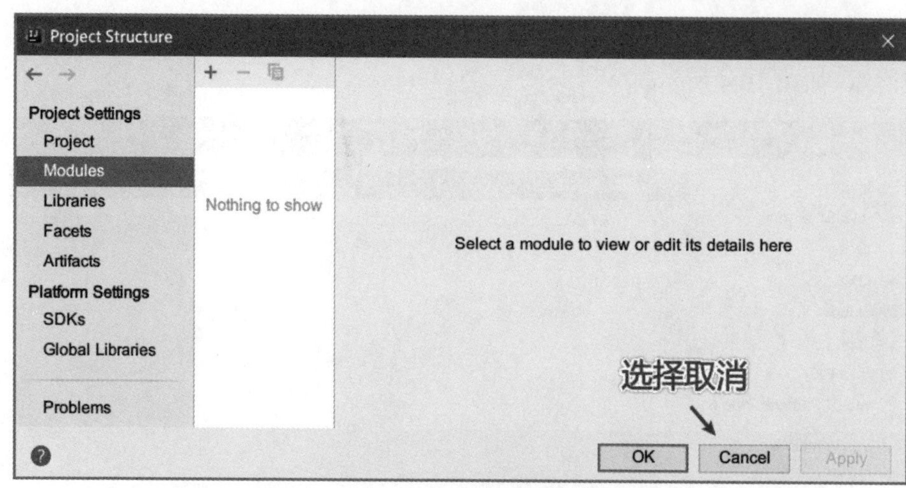

图 10-6　不创建模块

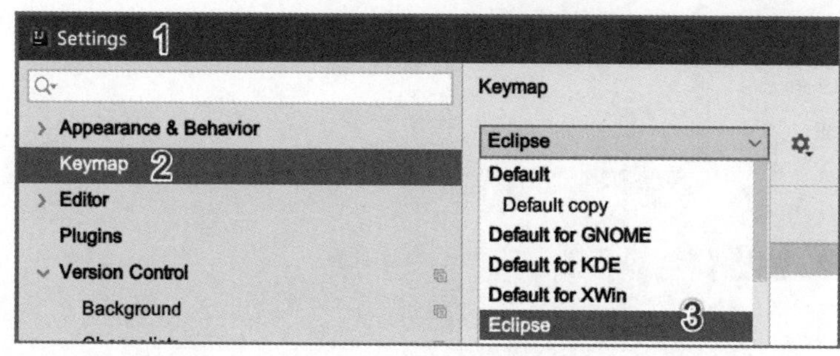

图 10-7　选择快捷键

（6）如果不进行快捷键的设置，那么 IDEA 常用的默认快捷键如表 10-1 所示。

表 10-1　IDEA 常用的快捷键

快捷键	功　能
Ctrl+Y	删除光标所在行
Ctrl+D	复制光标所在行并把复制的内容插入光标位置下面
Ctrl+Alt+L	格式化代码
Alt+Enter	导入包和异常处理等，自动修正代码，类似于 Eclipse 中的 Ctrl+1
Shift+Enter	在当前行的下面插入一个空行
Ctrl+Shift+Enter	在当前行的末尾插入一个分号
Ctrl+/	注释/反注释
双击 Shift	出现搜索对话框，可以搜索任何东西
Ctrl+Shift+/	选中代码注释，多行注释，再按取消注释
Ctrl+Space	自动补全代码（与中文输入法有冲突，建议修改成与 Eclipse 一样的 Alt+/）
Alt+Insert	自动生成代码，如 toString()、get()和 set()等方法
Ctrl + J	显示代码模板
Ctrl + P	提示方法参数
Ctrl + Alt + T	先选中代码块使用代码包围，如 if/else、try/catch 等

续表

快捷键	功　　能
Shift + F6	方法、变量的重命名
Ctrl + F12	在一个类中快速定位到指定的方法
Ctrl +Alt+O	优化导入的类和包

10.3 Vue 快速入门

视频讲解

1. 入门案例

下面通过一个案例快速体验 Vue 的使用，该案例的需求如下。

（1）在页面上显示姓名和住址的信息，如图 10-8 所示。

（2）单击“打招呼”按钮，弹出信息框，显示用户信息。

（3）单击“换地址”按钮，更换网页上姓名和住址的信息。

姓名：牛魔王
住址：火焰山
打招呼　换地址

图 10-8　Vue 入门案例效果 1

2. 创建模块

（1）选择 File 菜单 → New 选项 → Module 选项，创建一个模块。

（2）选择 Static Web 选项（在 2020 版的 IDEA 中，此处为 JavaScript），创建静态模块，如图 10-9 所示。

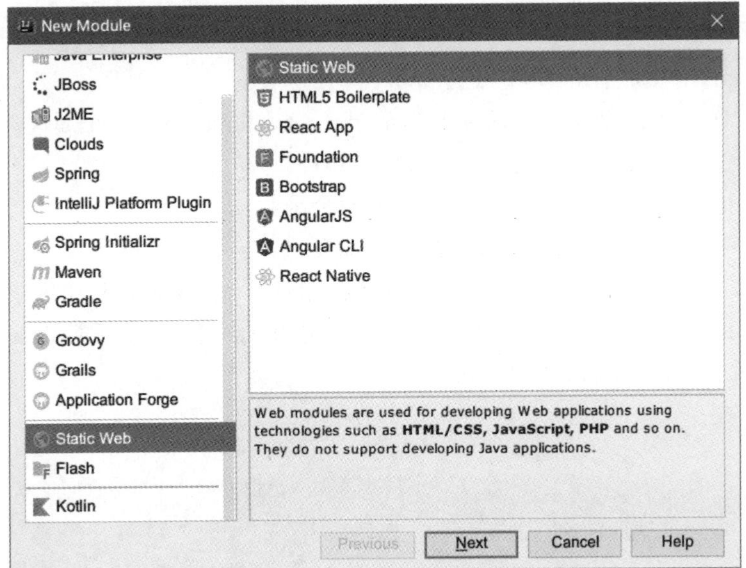

图 10-9　创建静态模块

（3）创建的模块结构如图 10-10 所示，vue.js 文件在书籍的配套资料中已经提供，读者也可以自行下载。在图 10-10 中，demo01.html 是新创建的，在 vue.js 的上一级目录，与 js 目录同级。

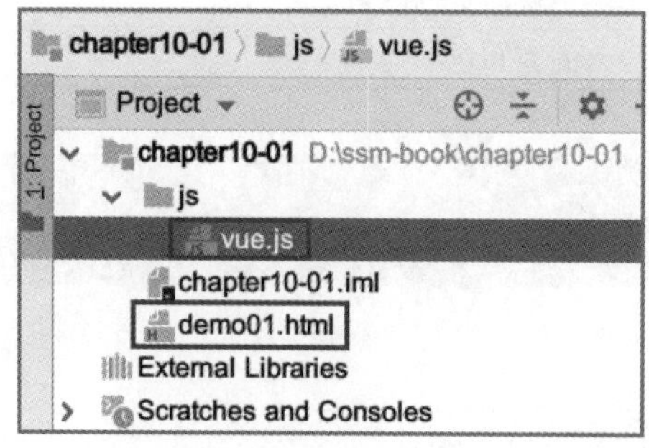

图 10-10　模块结构 1

（4）本书中的 JavaScript 是基于 ECMAScript6 的版本，因此需要先在 IDEA 中进行设置，否则开发 JS 会提示语法错误。选择 File 菜单→ Settings 选项→ Languages & Frameworks 选项→ JavaScript 选项，将语言版本设置为 ECMAScript6，如图 10-11 所示。

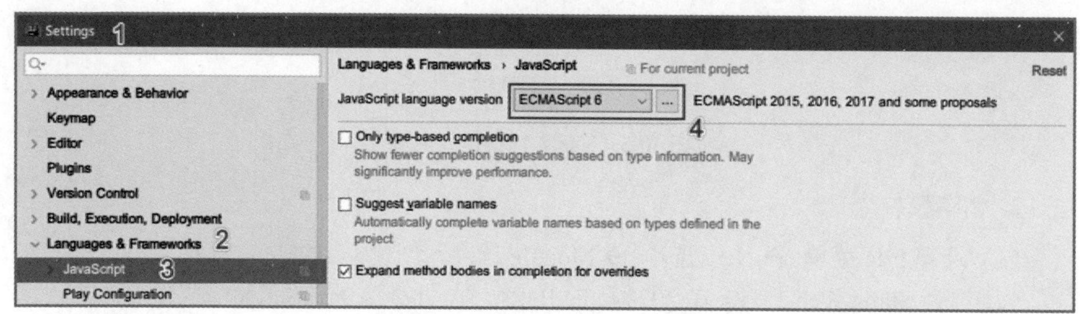

图 10-11　配置 ECMAScript6 环境

3. 开发步骤

（1）创建 div 作为视图，设置 id 为 app。

（2）编写 JS 脚本，创建 Vue 对象，指定 el、data 和 methods 三个属性。

① data：包含 name 和 address 两个属性，并且设置初始值。

② methods：包含 sayHi() 和 updateInfo()两个函数。

（3）在 div 中引用 name 和 address 两个插值变量。

（4）在 div 中再创建两个按钮，分别添加单击事件，调用 methods 中的事件处理函数，修改 name 和 address 的值。说明：在函数内部使用"this.属性"可以获取 data 中的属性值，给"this.属性"赋值可以修改 data 中的属性值。

4. 案例代码

```
<!DOCTYPE html>
<html lang="en">
<head>
    <meta charset="UTF-8">
```

```
    <title>快速体验：入门案例</title>
    <!-- 导入 vue.js -->
    <script src="js/vue.js"></script>
</head>
<body>
<div id="app">
    <!-- 引用 data 属性中的值，语法：{{变量名}}-->
    姓名：{{name}} <br/>
    住址：{{address}} <hr/>
    <!-- 单击事件调用函数，@click 表示单击事件 -->
    <button type="button" @click="sayHi()">打招呼</button>
    <button type="button" @click="updateInfo()">换地址</button>
</div>

<script type="text/javascript">
    /*
    Vue 对象使用三个属性：el, data, methods
    */
    new Vue({
        el: "#app",
        data: {
            name: "牛魔王",
            address: "火焰山"
        },
        methods: {
            sayHi() {
                //this 表示当前 Vue 对象，直接获取它的 data 中的属性值
                alert("你好，我是" + this.name + "，住在" + this.address);
            },
            updateInfo() {
                //直接修改 data 中的属性值
                this.name = "狐狸精";
                this.address = "狐狸洞";
            }
        }
    });

</script>
</body>
</html>
```

单击右上角的浏览器按钮即可直接运行，如图 10-12 所示。

图 10-12　浏览器按钮

可以尝试分别单击两个按钮，效果如图 10-13 所示。

图 10-13　Vue 入门案例效果 2

视频讲解

10.4　Vue 常用指令

1. 在 IDEA 中设置 Vue 代码提示

工欲善其事，必先利其器。先把 IDEA 工具打磨一下，以提高开发效率。按以下步骤进行操作：

（1）安装 IDEA 中支持 Vue 的插件，选择 File 菜单→Settings 选项→Plugins 选项，在 Marketplace 中搜索 Vue.js 进行安装，如图 10-14 所示。

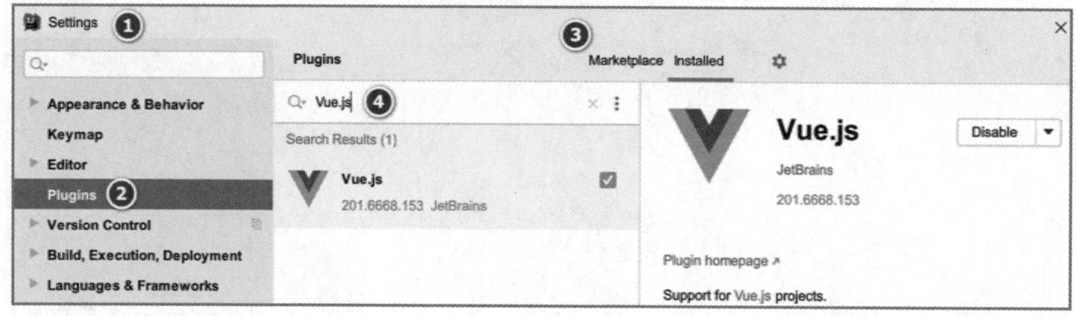

图 10-14　安装插件

（2）选择 File 菜单→Settings 选项→Editor 选项→File Types 选项→HTML 选项，再单击右边的+，在 HTML 中添加*.vue，如图 10-15 所示。

（3）在 HTML 中输入代码，可以看到 Vue 的指令提示。

2. 指令介绍

指令是指带有“v-前缀”的特殊属性，它可以告诉 Vue 如何将模型数据渲染在 HTML 的 DOM 树中，生成不同的网页结果并显示在浏览器中。指令通常作为属性名写在开始标签上，它的值支持 JavaScript 的表达式。

为了让 IDEA 中有 Vue 的语法提示，可以进行下面的设置。

3. 文本插值及案例

文本插值有两个指令，如表 10-2 所示。

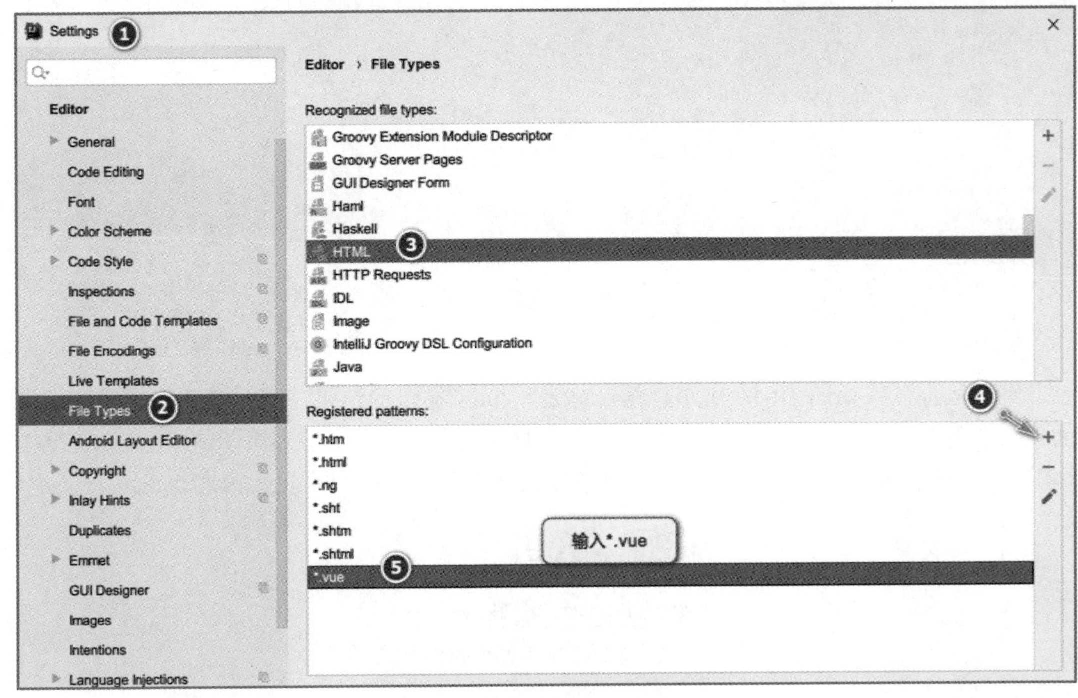

图 10-15 设置 HTML 文件提示

表 10-2 文本插值指令及说明

指 令	功 能 说 明
v-html	把属性值中的文本解析为 HTML，内容中出现 HTML 标签是起作用的
v-text	把属性值中的文本解析为纯文本，内容中出现 HTML 标签是不起作用的

案例实现步骤如下。

(1) 创建 2 个 div，在 div 上分别使用 v-html 和 v-text 指令。

(2) 输出 Vue 对象 data 属性中的 msg 值，msg 属性是一个带 html 标签的文本。

(3) 比较两个指令的区别。

案例代码如下。

```
<!DOCTYPE html>
<html lang="en">
<head>
    <meta charset="UTF-8">
    <title>v-html 和 v-text</title>
    <script src="js/vue.js"></script>
</head>
<body>
<div id="app">
    <div v-html="msg"></div>
    <div v-text="msg"></div>
</div>
```

```
    <script type="text/javascript">
        new Vue({
            el: "#app",
            data: {
                msg: "<h1 style='color:red'>你好，Vue!</h1>"
            }
        });
    </script>
    </body>
    </html>
```

案例运行效果如图 10-16 所示。第一行是 v-html 输出的结果，其标签和样式是有效果的；第 2 行是 v-text 输出的结果，其标签是原样输出的。为了避免因用户输入一些 HTML 标签或 JS 脚本导致的 HTML 注入，有些论坛应该使用 v-text 显示页面内容。

你好，Vue!

<h1 style='color:red'>你好，Vue!</h1>

图 10-16　文本插值效果

视频讲解

10.5　绑定属性

绑定属性的指令及说明如表 10-3 所示。

表 10-3　绑定属性的指令及说明

指　　令	功 能 说 明
v-bind:属性名="值"	为 HTML 的普通属性绑定值，值指定为 data 中的变量名。当变量发生变化时，标签的属性值也发生变化，也就是模型发生变化会影响视图的变化
:属性名="值"	这是另一种简写的方式，功能同上

1. 案例步骤

（1）给 a 标签的 href 设置属性值，取值来自 data 中的变量名。

（2）使用 v-bind:属性名="data 中属性名"指令绑定一个变量。注意：如果出现红色提示，按 Alt+Enter 键添加一个 HTML 命名空间即可。

（3）使用:href="data 中属性名"实现同样的功能。

2. 案例代码

```
<!DOCTYPE html>
<html lang="en" xmlns:v-bind="http://www.w3.org/1999/xhtml">
<head>
    <meta charset="UTF-8">
    <title>v-bind绑定属性值</title>
```

```
    <script src="js/vue.js"></script>
</head>
<body>
<div id="app">
    <!-- 使用 v-bind 绑定属性值-->
    <a v-bind:href="url">百度</a><br/>
    <!-- 简写-->
    <a :href="url">百度</a><br/>
</div>

<script type="text/javascript">
    new Vue({
        el: "#app",
        data: {
            url: "http://www.baidu.cn"
        }
    });
</script>
</body>
</html>
```

3. 案例效果

在 Chrome 浏览器中按 F12 键，选择 Elements 可以看到浏览器中最终渲染出来的网页元素，渲染效果如图 10-17 所示。

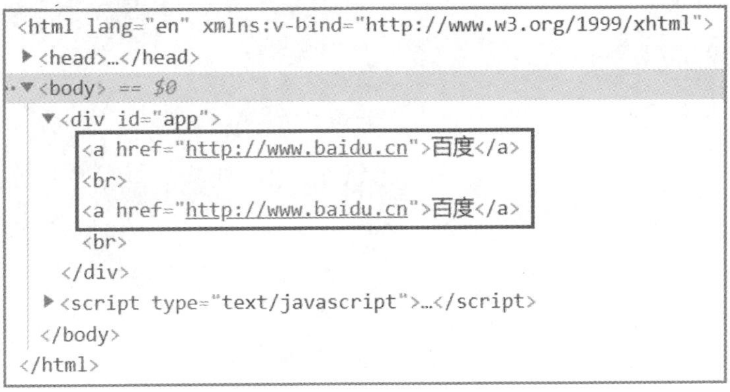

```
<html lang="en" xmlns:v-bind="http://www.w3.org/1999/xhtml">
 ▶<head>…</head>
 ▼<body> == $0
   ▼<div id="app">
      <a href="http://www.baidu.cn">百度</a>
      <br>
      <a href="http://www.baidu.cn">百度</a>
      <br>
    </div>
   ▶<script type="text/javascript">…</script>
  </body>
</html>
```

图 10-17　浏览器渲染结果

这两个链接是可以单击的，单击后会跳转到百度的网站，说明这两种写法的功能是一样的。

10.6 绑定事件

绑定事件指令及说明如表 10-4 所示。

表 10-4　绑定事件指令及说明

指　令	功能说明
v-on:事件名="处理函数"	为 HTML 标签绑定事件，调用 Vue 对象中 methods 的函数
@事件名="处理函数"	另一种写法，功能同上

1. 案例步骤

1）创建页面

（1）创建 HTML 页面，设置一个文本框和两个按钮。

（2）文本框的值绑定 data 中的 username 属性。

2）添加事件

（1）单击第 1 个按钮：使用语法一，编写单击事件，修改文本框的内容。

（2）单击第 2 个按钮：使用语法二，编写单击事件，修改文本框的内容。

注意： 如果方法没有参数，函数调用时的括号可以省略。

2. 案例代码

```
<!DOCTYPE html>
<html lang="en" xmlns:v-on="http://www.w3.org/1999/xhtml">
<head>
    <meta charset="UTF-8">
    <title>v-on 绑定事件</title>
    <script src="js/vue.js"></script>
</head>
<body>
<div id="app">
    <!-- 绑定 data 中的 username -->
    <input type="text":value="username">
    <hr/>
    <!-- 绑定 methods 中的函数名，如果调用没有参数，函数后面的括号可以省略 -- >
    <input type="button" v-on:click="change1()" value="变身 1">
    <input type="button" @click="change2()" value="变身 2">
</div>

<script type="text/javascript">
    new Vue({
        el: "#app",
        data: {
            username: "白骨精"
        },
        methods: {
            change1() {
                this.username = "孙悟空";
            },
            change2() {
                this.username = "猪八戒";
```

```
            }
        }
    });
</script>
</body>
</html>
```

3. 案例效果

单击不同的按钮，会修改文本框中的值，如图 10-18 所示。

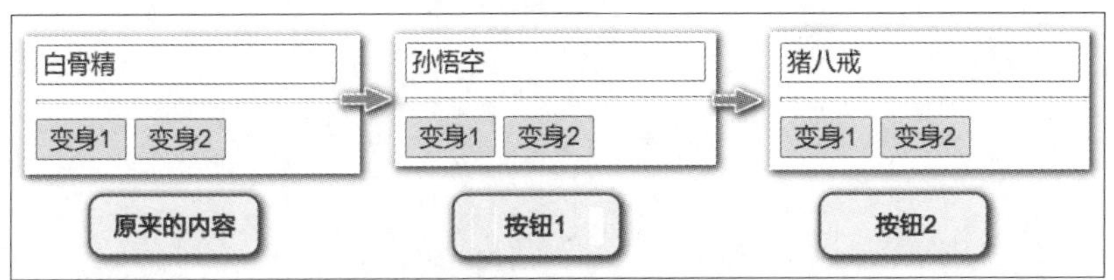

图 10-18 绑定事件案例效果

10.7 条件渲染

条件渲染指令及说明如表 10-5 所示。

表 10-5 条件渲染指令及说明

指　　令	功 能 说 明
v-if	有条件地渲染当前元素，如果判断结果为真就渲染到网页上，否则不进行渲染
v-else-if	充当 v-if 的 else-if 块，可以连续使用
v-else	必须紧跟在带 v-if 或者 v-else-if 元素的后面，否则它将不会被识别

1. 案例需求

根据 data 中 type 属性值进行判断，type 值等于 A、B、C 或其他，显示“优秀”、“良好”、“及格”或“不及格”所在的 div。这里的判断条件使用了 JS 表达式。

2. 案例代码

```
<!DOCTYPE html>
<html lang="en">
<head>
    <meta charset="UTF-8">
    <title>条件渲染</title>
    <script src="js/vue.js"></script>
</head>
```

```
<body>
<div id="app">
    <!-- v-if 中可以编写 JS 表达式，注：外面是双引号，里面的字符串用的单引号 -- >
    <div v-if="type=='A'">
        优秀
    </div>
    <div v-else-if="type=='B'">
        良好
    </div>
    <div v-else-if="type=='C'">
        通过
    </div>
    <div v-else>
        不通过
    </div>
</div>
<script type="text/javascript">
    let app = new Vue({
        el: "#app",
        data: {
            type: 'B'
        }
    });
</script>
</body>
</html>
```

3. 案例效果

网页最开始显示的内容是良好，app 是 Vue 对象在网页中的变量名，如图 10-19 所示。在浏览器中按 F12 键，可以在 Console 栏中动态地改变 type 的值。随着 type 值的变化，网页上渲染出来的 div 内容也是不同的，这是因为模型的变化会导致视图的变化。

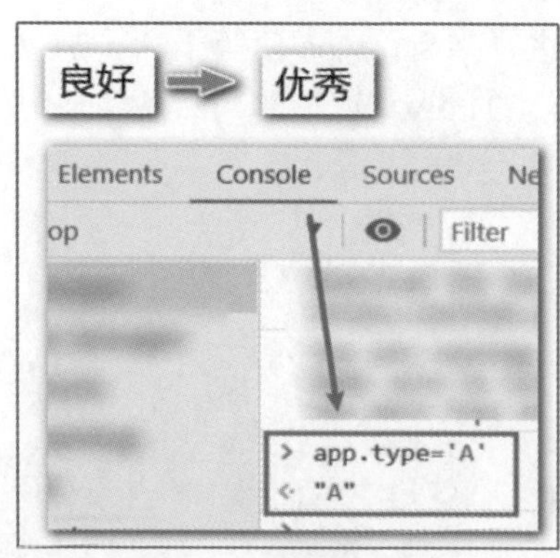

图 10-19　条件渲染案例效果

10.8 循环渲染

循环渲染指令及说明如表 10-6 所示。

表 10-6　循环渲染指令及说明

指　令	功 能 说 明
v-for	列表渲染：用于遍历数组中的每个元素，遍历对象的每个属性，也可以用于固定循环次数

1. 遍历的语法

- 遍历普通数组的语法。

```
v-for="变量名 in 数组"
输出变量名使用：{{变量名}}
```

- 遍历对象数组的语法。

```
v-for="(属性值,属性名,索引) in 对象"
注意：可以只写前面 1 个或 2 个变量。
```

- 遍历固定循环次数的语法。

```
v-for="变量名 in 整数"
也可以接受一个整数值，遍历一个数值范围，类似于 Java 中的 for-i 循环，起始值从 1 开始。
```

2. 案例代码

```
<!DOCTYPE html>
<html lang="en">
<head>
    <meta charset="UTF-8">
    <title>v-for 遍历</title>
    <script src="js/vue.js"></script>
</head>
<body>
<div id="app">
    <ul>
        <!-- 遍历普通数组-->
        <li v-for="name in names">{{name}}</li>
    </ul>

    <ul>
        <!-- 遍历对象数组-->
        <li v-for="(person, index) in persons">第{{index}}号 姓名：{{person.name}}
年龄：{{person.age}}</li>
```

```
        </ul>

        <ul>
            <!--遍历对象的属性-->
            <li v-for="(value,key) in person">属性值：{{value}}， 属性名：{{key}}
</li>
        </ul>

        <!-- 遍历固定次数 -->
        <span v-for="I in 10">{{i}} </span>
    </div>

    <script type="text/javascript">
        new Vue({
            el: "#app",
            data: {
                //普通数组
                names: ["Jack","Rose","NewBoy","Tom"],
                //对象数组
                persons: [{
                    name: "孙悟空",
                    age: 30
                },{
                    name: "猪八戒",
                    age: 28
                },{
                    name: "沙悟净",
                    age: 26
                }],
                //对象
                person : {
                    name: "唐三藏"
                    age: 20,
                    sex: "男"
                }
            }
        });
    </script>
    </body>
    </html>
```

3. 案例效果

对于普通数组，输出数组中每个字符串的值；对于对象数组，依次输出每个对象的索引号和属性值；对于对象属性，依次输出对象的每个属性名和属性值。固定次数则类似于for-i 循环，输出每次累加的值，起始值从 1 开始。

循环渲染效果如图 10-20 所示。

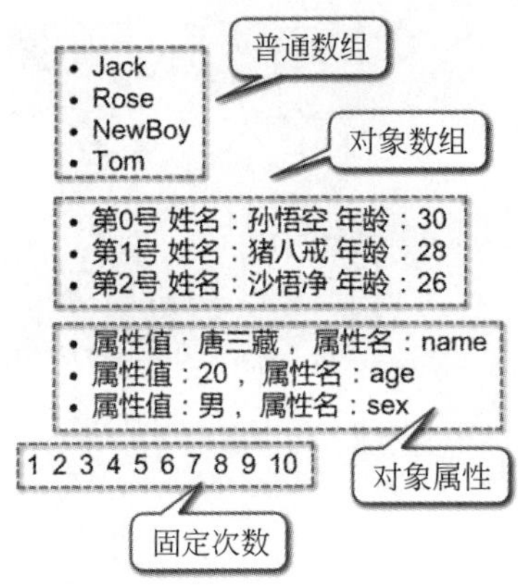

图 10-20　循环渲染案例效果

10.9 双向绑定

1. 关于数据绑定的说明

双向绑定用于表单元素中，即出现在 form 容器中的元素，如 input、select 和 textarea 等。

- 表单绑定

v-model：在表单元素上创建双向数据绑定。

- 单向数据绑定

数据 Model 发生改变→视图 View 内容发生改变，但视图 View 的内容改变不会造成数据 Model 改变。

- 双向数据绑定

数据 Model 发生改变→视图 View 内容会改变。

视图 View 的内容改变→数据 Model 也会改变。

2. MVVM 模式介绍

Vue 提供了 MVVM 风格的双向数据绑定的功能，专注于 View 层。MVVM 模型（Model，View，ViewModel）是 MVC 模式的改进版，它的核心是 MVVM 中的 VM，也就是 ViewModel。ViewModel 负责连接 View 和 Model，保证视图和数据的一致性。

在前端页面中，JS 对象表示 Model，页面表示 View，两者最大限度地分离。将 Model 和 View 关联起来的就是 ViewModel，它相当于一座桥梁。ViewModel 负责把 Model 的数据在 View 中同步显示出来，并负责把 View 修改的数据同步回 Model 中。MVVM 模型如图 10-21 所示。

3. 案例步骤

（1）显示用户信息页面，给表单中每一项 v-model 绑定 data 中对应的属性名。

（2）在 methods 的方法中创建 showInfo 方法，给按钮添加单击事件绑定 showInfo。

（3）在 showInfo 中输出每一项的属性值。

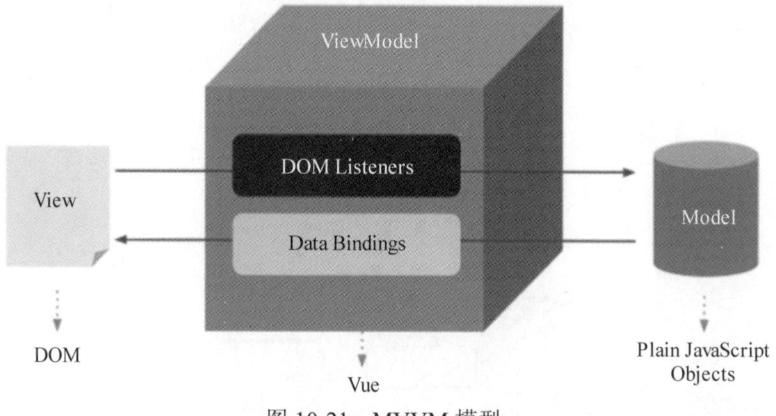

图 10-21　MVVM 模型

4. 案例代码

```
<!DOCTYPE html>
<html lang="en">
<head>
    <meta charset="UTF-8">
    <title>v-model 双向绑定值</title>
    <script src="js/vue.js"></script>
</head>
<body>
<div id="app">
    姓名：<input type="text" name="name" v-model="name"> <br>
    性别：
    <input type="radio" name="sex" v-model="sex" value="男">男
    <input type="radio" name="sex" v-model="sex" value="女">女
    <br>
    爱好：
    <input type="checkbox" name="hobby" value="游泳" v-model="hobby">游泳
    <input type="checkbox" name="hobby" value="看书" v-model="hobby">看书
    <input type="checkbox" name="hobby" value="下棋" v-model="hobby">下棋
    <br>
    学历：
    <select name="edu" v-model="edu">
        <option>高中</option>
        <option>大专</option>
        <option>本科</option>
        <option>硕士</option>
    </select>
    <hr>
    <input type="button" value="显示值" id="btnBind" @click="showInfo">
</div>

<script type="text/javascript">
    let app = new Vue({
```

```
        el: "#app",
        data: {
            name: "白骨精",
            sex: "女",
            hobby:["看书","下棋"],
            edu: "本科"
        },
        methods: {
            showInfo() {
                alert("姓名：" + this.name + "，性别：" + this.sex + "，爱好：" + this.hobby + "，学历：" + this.edu);
            }
        }
    });
</script>
</body>
</html>
```

5. 案例测试

（1）每修改一次表单项的值（View），单击按钮查看一下 data 的值（Model），这是 View 影响 Model。

（2）在浏览器的 Console 控制台修改属性值（Model），查看表单项的内容是否变化（View），这是 Model 影响 View。

双向绑定效果如图 10-22 所示。

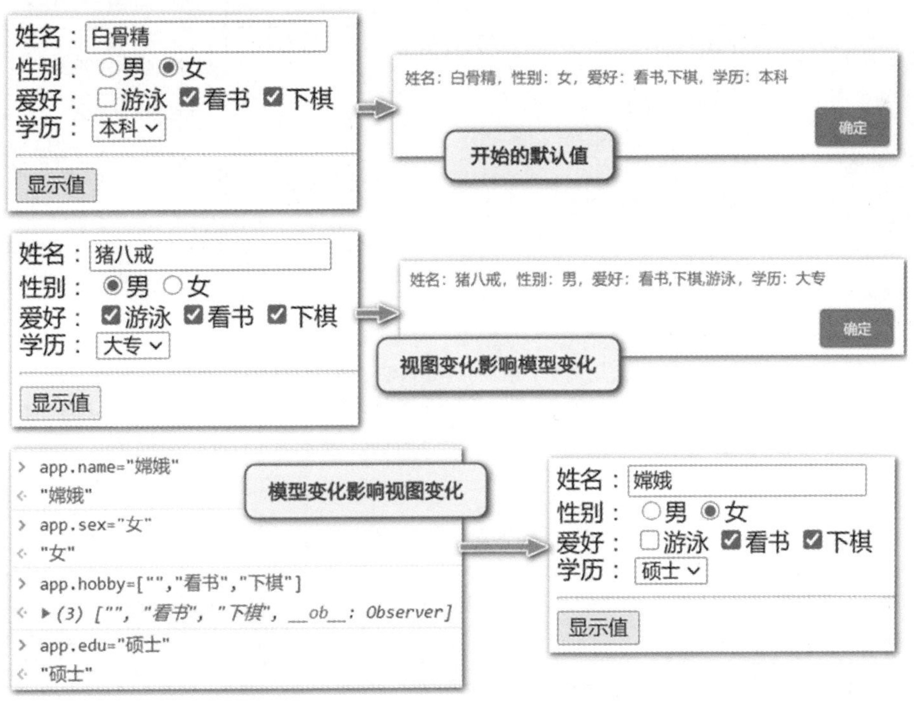

图 10-22　双向绑定测试运行效果

视频讲解

10.10 Vue 的 AJAX 异步操作

1. Axios 介绍

Vue 本身不支持发送 AJAX 请求，需要使用 vue-resource、Axios 等插件来实现。Axios 是一个基于 Promise 的 HTTP 请求客户端，用来发送请求，也是 Vue2.0 官方推荐的插件。

Axios 的下载地址为 https://github.com/axios/axios，详细的使用可以查看它的 README.md 文档，如图 10-23 所示。

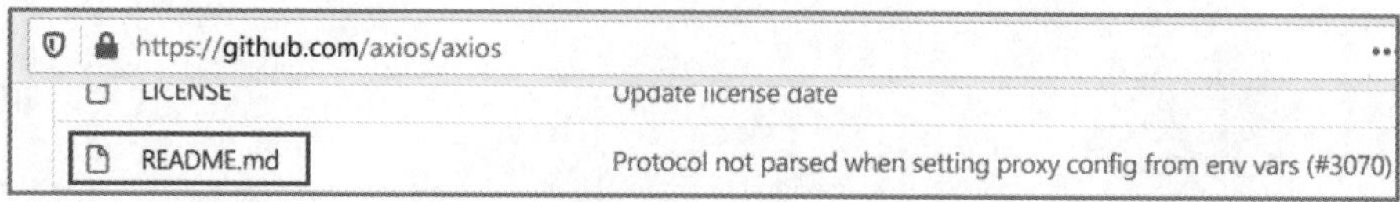

图 10-23　Axios 的仓库

Axios 的使用流程如下。

（1）引入 Axios 核心 js 文件。

（2）调用 Axios 对象的方法来发起异步请求。

（3）调用 Axios 对象的方法来处理响应的数据。

2. Axios 的方法和语法（表 10-7）

表 10-7　Axios 的语法

Axios 对象的方法	方 法 说 明	举　　例
get('url?参数')	GET 请求地址和参数，参数直接写在地址后面。 参数格式：键=值&键=值	axios.get('/user?ID=12345')
post('url',参数)	POST 请求的地址和参数，参数与地址要分开写。 参数格式 1：键=值&键=值 参数格式 2：{键:值,键:值} 注意：参数格式 2 需要在服务器端进行转换处理	axios.post('/user', { 　firstName: 'Boy', 　lastName: 'New' })

then(正常回调函数)	回调函数的参数是服务器返回的对象，包含以下属性。
	data：表示服务器返回的数据。
	status：表示服务器的状态码。
	注意：如果使用匿名的回调函数写法，在函数体内的 this 不是 Vue 对象。但用 ES6 的箭头语法，则可以使用 this 来引用 Vue 中的 data 数据
		.then(function (response) {
		　console.log(response);
		})
		或
		.then(response => {
		　console.log(response);
		})

catch(异常回调函数)	回调函数的参数是服务器返回的错误对象，包含以下属性。
	message：服务器返回的错误信息
		.catch(function (error) {
		　console.log(error);
		})
		或
		.catch(error => {
		　console.log(error);
		})

3. 案例需求

（1）模拟用户注册时判断用户名是否存在。

（2）使用 GET 或 POST 请求提交用户值。

（3）在服务器端使用一个 Servlet 来处理业务。

（4）假设用户名为 newBoy，则表示用户已经存在，否则都可以注册。

4. 案例步骤

Axios 的模块结构如图 10-24 所示。

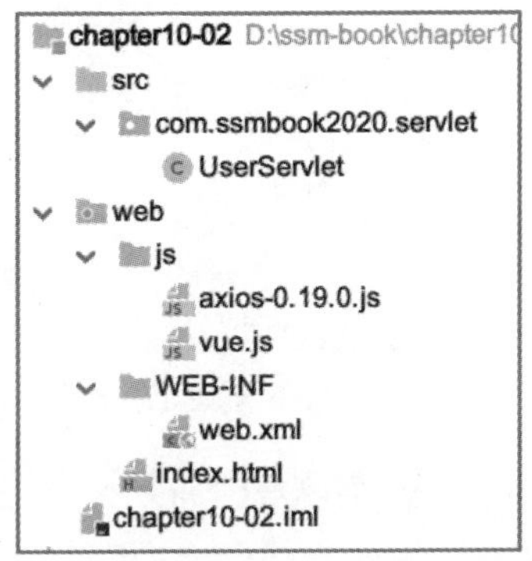

图 10-24　Axios 的模块结构

（1）在 IDEA 中创建一个新的 Web 模块，选择 File 菜单→New 选项→Module 选项。

（2）配置 Web 模块的各项参数，如图 10-25 所示。

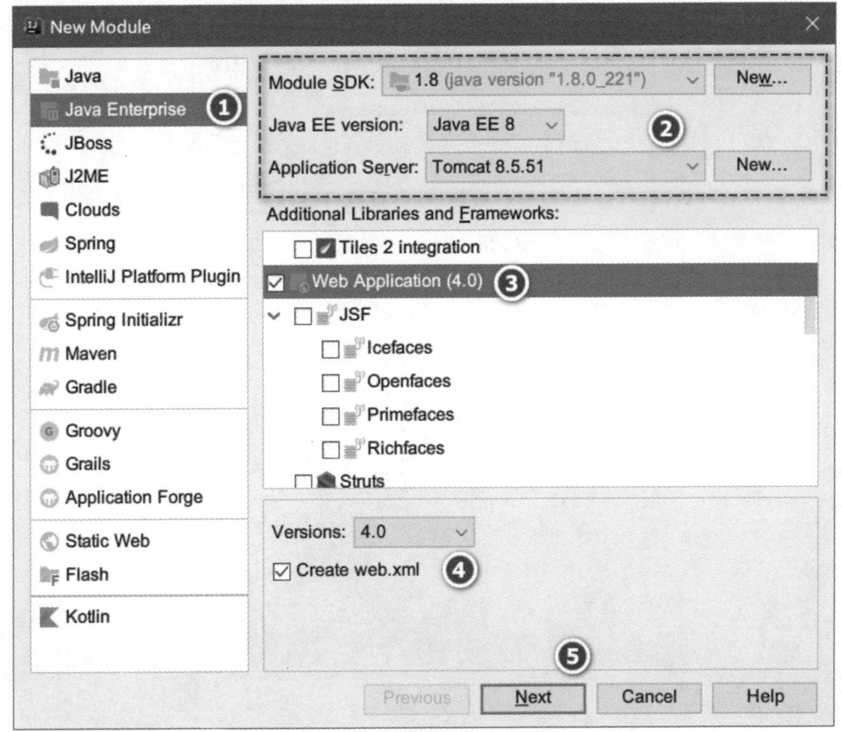

图 10-25　Web 参数设置

（3）指定模块的名字和位置，如图 10-26 所示。

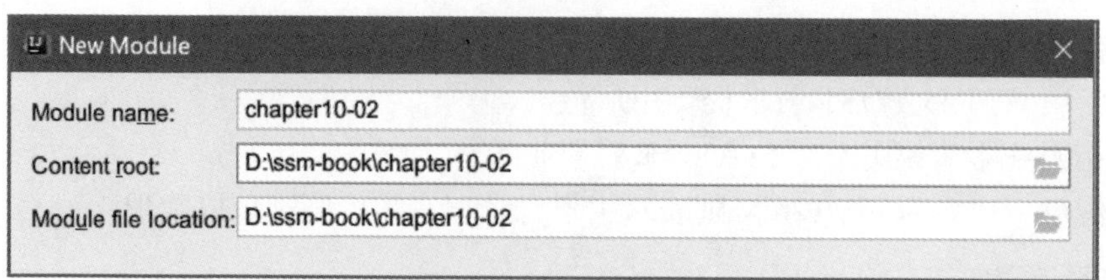

图 10-26　模块名字

（4）创建的工程结构如图 10-27 所示。

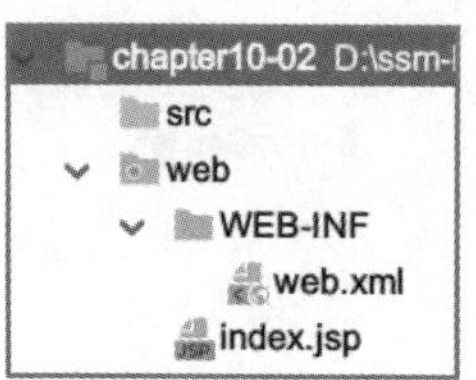

图 10-27　Web 模块工程结构

（5）在 src 目录上右击，选择 New 菜单→Create New Servlet 选项，创建一个 Servlet，参数设置如图 10-28 所示。

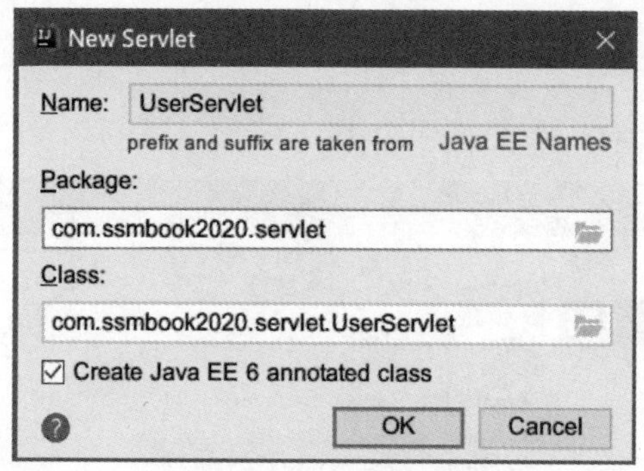

图 10-28　Servlet 参数设置

Servlet 代码如下。

```
Package com.ssmbook2020.servlet;

import javax.servlet.ServletException;
import javax.servlet.annotation.WebServlet;
import javax.servlet.http.HttpServlet;
import javax.servlet.http.HttpServletRequest;
import javax.servlet.http.HttpServletResponse;
import java.io.IOException;
import java.io.PrintWriter;

//Servlet 的访问地址
@WebServlet("/user")
```

```
public class UserServlet extends HttpServlet {
    protected void doPost(HttpServletRequest request, HttpServletResponse
response) throws ServletException, IOException {
        //设置响应的结果
        response.setContentType("text/plain;charset=utf-8");
        PrintWriter out = response.getWriter();
        //POST 有汉字乱码的问题
        request.setCharacterEncoding("utf-8");
        //获取客户端提交的用户名参数
        String username = request.getParameter("username");
        //在服务器端打印出来
        System.out.println("请求的方式是: " + request.getMethod());
        System.out.println("用户名是: " + username);
        //判断用户是否存在
        if ("newboy".equalsIgnoreCase(username)) {
            out.print("用户已经存在");
        }
        else {
            out.print("恭喜可以注册");
        }
    }

    protected void doGet(HttpServletRequest request, HttpServletResponse
response) throws ServletException, IOException {
        this.doPost(request, response);
    }
}
```

（6）HTML 页面的代码要同时导入 vue 和 axios 两个 js 文件。

```
<!DOCTYPE html>
<html lang="zh-CN">
<head>
    <meta charset="UTF-8">
    <title>注册：判断用户是否存在</title>
    <!-- 导入 vue -->
    <script src="js/vue.js"></script>
    <!--导入 axios -->
    <script src="js/axios-0.19.0.js"></script>
</head>
<body>
<div id="app">
    <!-- 失去焦点判断用户名是否存在 -->
    用户名：<input type="text" name="username" id="username" @blur=
"checkUser()">
    <!-- 显示判断的结果 -->
    <span id="userInfo">{{msg}}</span>
```

```
    </div>

    <script type="text/javascript">
        new Vue({
            el: "#app",
            data: {
                //默认 msg 中的内容为空
                msg: "",
            },
            methods: {
                checkUser() {
                    //请求的地址和参数
                    let param = "username=" + document.getElementById("username").
value;
                    //POST 请求
                    axios.post("user", param)
                        .then(result => {
                        //将服务器返回的结果直接给 msg 值，显示即可
                        this.msg = result.data;
                    }).catch(error => {
                        alert("服务器出现异常：" + error.message);
                    });
                }
            }
        });
    </script>
    </body>
    </html>
```

5. 部署到 Web 容器运行

（1）选择 Run 选项→Edit Configurations 选项，在弹出的窗口中，将当前的模块部署到 Tomcat，如图 10-29 所示。

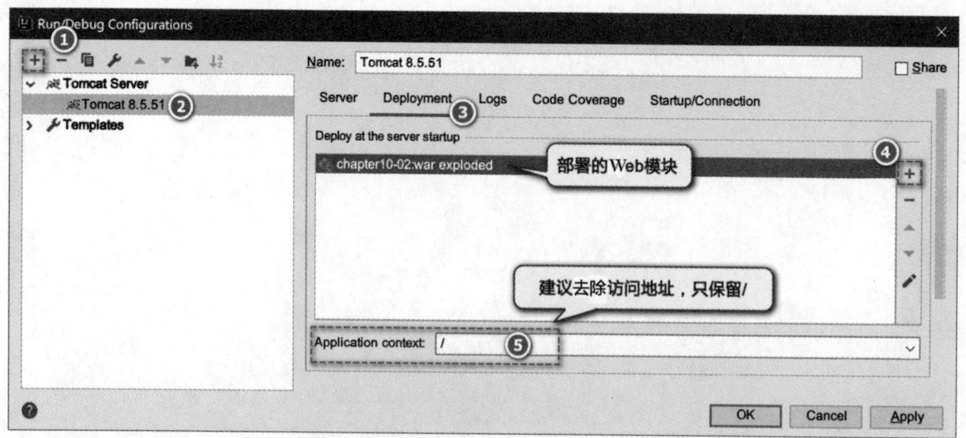

图 10-29　部署到 Tomcat

（2）单击工具栏上的运行按钮运行。

分别在文本框中输入张三和 newboy，可以在服务器端看到输出结果。

```
请求的方式是：POST
用户名是：张三
请求的方式是：POST
用户名是：newboy
```

（3）在浏览器上可以看到输出结果，AJAX 异步操作效果如图 10-30 所示。

图 10-30　AJAX 异步操作案例效果

10.11 本章小结

本章首先介绍了 Vue 前端框架概念及其组成和特点，它是一种 MVVM 架构的前端开发框架。然后介绍了 IntelliJ IDEA 开发工具的使用，这是一个比 MyEclipse 更优秀的商用集成开发环境。最后通过一个入门案例快速了解了 Vue 的基本开发结构。

本章重点介绍了 Vue 常用的指令，其中包括文本插值、绑定属性、绑定事件、条件渲染、循环渲染和双向绑定。对于后端 Java 程序员来说，接触比较多的是表单元素，所以双向绑定是比较常用的。

此外，本章还讲解了 Vue 中的 AJAX 异步操作，这需要通过第三方插件 Axios 来实现。

习题 10

参考 10.10 节 Vue AJAX 异步操作的相关内容，实现一个用户登录的功能。如果登录成功就在页面显示登录成功的信息，如果登录失败就显示失败的信息。

用户登录功能的需求如下。

（1）页面上有用户名和密码两个文本框，单击“登录”按钮，使用后台 AJAX 完成登录操作。

（2）如果用户登录成功则显示登录成功的信息，否则显示登录失败的信息。

（3）如果用户名为 newboy 且密码为 123，则登录成功，后台服务器暂不使用数据库。

用户登录功能的效果应该如图 10-31 所示。

图 10-31　用户登录功能的效果

第 11 章 Element+SSM 开发员工管理模块

本章学习内容

- Maven 基础；
- 使用 Maven 搭建 SSM 环境；
- 员工管理系统的实现；
- 基于 Element 框架的系统开发。

11.1 Maven 基础

11.1.1 为什么要学习 Maven

在没有使用 Maven 之前，在开发过程中有很多痛苦的开发经历，通常会有以下几种。

（1）包依赖的问题：JAR 包一般都不是独立存在的，在使用时会用到其他的 JAR。例如，a.jar 依赖于 b.jar，而 b.jar 又依赖于 c.jar。当用到 a.jar 时，需要把 3 个 JAR 包都加载进来才可以使用。当项目中用到很多 JAR 包时，很难判断缺少哪些 JAR 包，只有在项目运行过程报错后才知道。

（2）包版本冲突问题：项目中用到了 a.jar，a.jar 依赖于 c.jar 的 1.5 版本，然后把这两个 JAR 复制到项目中。后来又用到了 b.jar，但是 b.jar 依赖于 c.jar 的 1.0 版本，需要把 b.jar 和 c.jar 的 1.0 版本引进来。这时会发现 c.jar 有两个版本，发生了冲突。这种情况要解决 JAR 包冲突的问题，也是非常痛苦的。

（3）包的管理不方便：当项目比较大时，会将一个大的项目分成很多小的项目，每个小项目由几个开发者负责，然后每个小项目都需要把这些 JAR 复制一份到自己的项目目录中。

（4）项目开发结构不统一：Maven 可以让不同的项目按照某种规范采用相同的结构，这样会很方便。

（5）项目生命周期的控制流程复杂：开发者除了编码之外，多数时间都在重复着单元测试、编译、打包、发布的工作。在没有自动化编译的时候，每个过程都需要手动操作。

Maven 的出现可以很好地解决上面这些问题。

11.1.2 Maven 基本概念

1. Maven 仓库

Maven 将所有的 JAR 包都放到仓库里面，在 pom.xml 文件中通过坐标就可以将仓库里的 JAR 包引用到项目中。Maven 仓库的分类如表 11-1 所示。

表 11-1 Maven 仓库的分类

仓库名称	作　用
本地仓库	相当于缓存，工程第一次会从远程仓库（互联网）下载 JAR 包，将 JAR 包存在本地仓库（在本地电脑上）。以后就不需要从远程仓库下载，它会先从本地仓库找，如果找不到才去远程仓库找

续表

仓库名称	作　　用
远程仓库 (分成三种)	1. 中央仓库：JAR 包由专业团队（Maven 团队）统一维护，有全球最完整、最通用的 JAR 包。中央仓库的访问地址为 https://repo1.maven.org/maven2/。 2. 第三方仓库：由大型公司搭建的供使用的 JAR 包服务器，例如，阿里云仓库 https://maven.aliyun.com/mvn/guide。 3. 私服：在公司内部架设 JAR 包的私有服务器，访问私服仓库比访问中央仓库速度更快、更稳定

2. Maven 坐标

Maven 的核心作用是管理项目的依赖和引入所需的各种 JAR 包。为了能自动地解析任何一个 JAR 包，Maven 必须将这些 JAR 包或者其他资源进行唯一标识，即 Maven 坐标。Maven 坐标由一组元素定义，它们可以用来唯一标识项目、依赖、插件等。如何定义坐标元素？它有 groupId、artifactId 和 version 三个关键属性，每个属性的作用可以参考表 11-2。

表 11-2　坐标元素的定义

元素名称	说　　明
groupId	团体、公司或组织的名字，通常它以创建这个项目的组织名称的逆向域名开头，类似于 Java 中的包名
artifactId	项目名或模块名，表示一个单独项目或者模块的唯一标识符
version	项目的特定版本，正在开发中的项目可以用 SNAPSHOT 加上一个特殊的标记

例如，要引入 JUnit 的 JAR 包，只需要在 pom.xml 配置文件中配置引入 JUnit 的坐标即可，示例如图 11-1 所示。

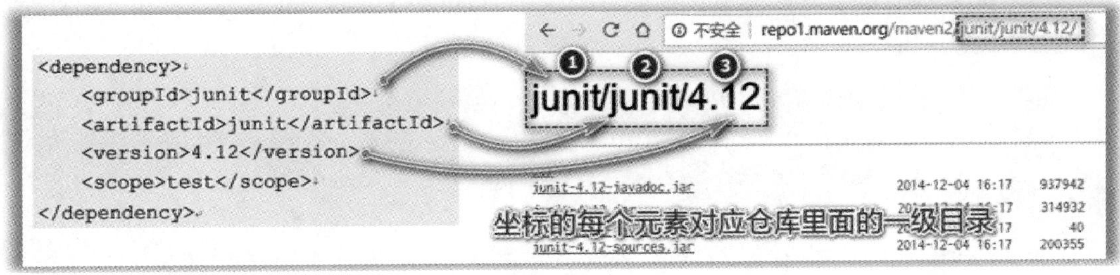

图 11-1　坐标示例

通过以上三层结构就可以在仓库中定位到一个 JAR 包，坐标的每个组成部分都对应仓库里面的一级目录结构，在本地仓库中也可以找到。

有时 groupId 可能对应多级目录，但 artifactId 和 version 只对应一级目录。

11.1.3　Maven 的安装与配置

视频讲解

Maven 的安装与配置步骤如下。

1. 下载 Maven

Maven 官网地址为 http://maven.apache.org/ ，可以在这里进行下载，本书使用的是 3.5.2

版本。因为最新的 3.6.3 版本与 IDEA2019 年 3 月以前的版本不兼容，建议最好与本书的版本一致。Maven 官网首页如图 11-2 所示。

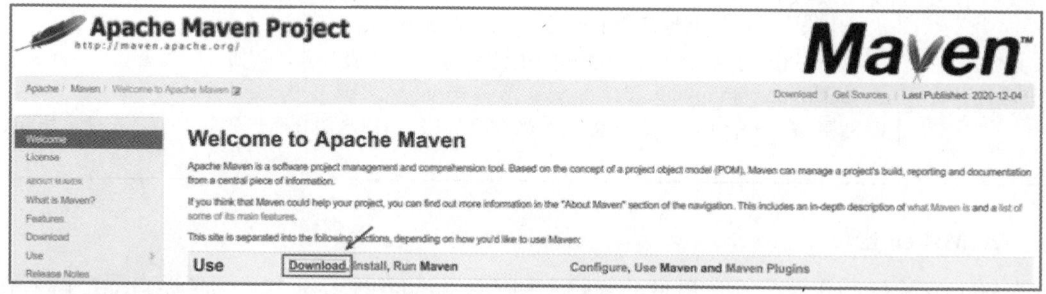

图 11-2　Maven 官网首页

2. 安装 Maven

将 Maven 压缩包 apache-maven-3.5.2-bin.zip 解压到任意目录即可，但需要注意以下几点。

（1）不要放在有汉字的目录下。

（2）目录层级不要太深。

（3）目录中尽量不要有空格。

3. 配置本地仓库

（1）在本机创建一个空目录，作为本地仓库。

```
e:\repository
```

（2）配置本地仓库，修改 Maven 安装目录中的 conf/settings.xml 文件，在第 53 行配置本地仓库为上面的空目录。

```
<settings xmlns="http://maven.apache.org/SETTINGS/1.0.0"
xmlns:xsi="http://www.w3.org/2001/XMLSchema-instance"
xsi:schemaLocation="http://maven.apache.org/SETTINGS/1.0.0
http://maven.apache.org/xsd/settings-1.0.0.xsd">
<localRepository>e:\repository</localRepository>
</settings>
```

（3）配置远程仓库，修改 settings.xml 文件，在第 146 行指定中央仓库的镜像。这里使用的是阿里云的仓库，速度比中央仓库快很多。注意：该部分代码要放在 mirrors 元素的下一级。

```
<mirror>
    <id>nexus-aliyun</id>
    <mirrorOf>central</mirrorOf>
    <name>Nexus aliyun</name>
    <url>http://maven.aliyun.com/nexus/content/groups/public</url>
</mirror>
```

（4）配置 JDK 版本，在 settings.xml 文件的第 189 行加入如下信息。注意：该部分代码要写在 profiles 元素下面。

```
<profile>
   <id>development</id>
   <activation>
     <jdk>1.8</jdk>
     <activeByDefault>true</activeByDefault>
   </activation>
   <properties>
     <maven.compiler.source>1.8</maven.compiler.source>
     <maven.compiler.target>1.8</maven.compiler.target>
     <maven.compiler.compilerVersion>1.8</maven.compiler.compilerVersion>
   </properties>
 </profile>
```

4. 配置环境变量与测试

如图 11-3 所示，右击"我的电脑"，选择"属性"→"高级系统设置"→"高级"，单击"环境变量"按钮。

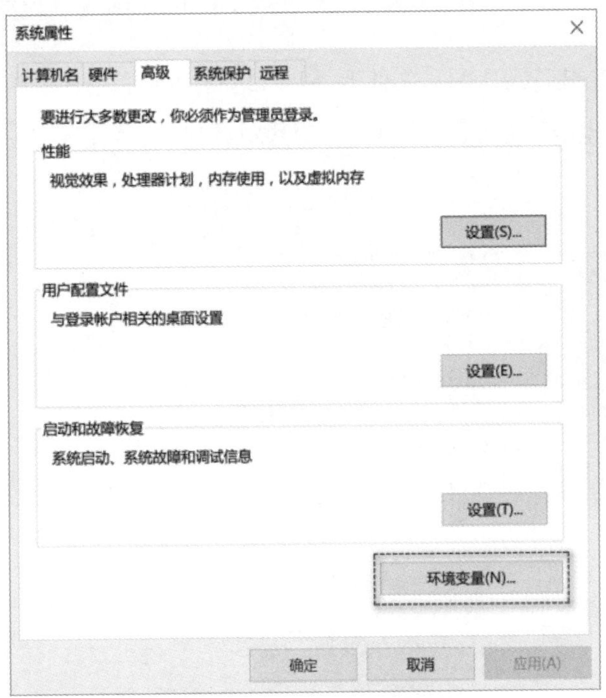

图 11-3　配置 Maven 环境变量

添加 Maven 的主目录为安装目录，设置 Maven 的可执行文件访问路径为 Maven 主目录的 bin 目录。

```
MAVEN_HOME=e:\apache-maven-3.5.2
Path=%MAVEN_HOME%\bin;
```

最后测试安装好的 Maven。打开 cmd 本地控制台，输入 mvn -v，若出现下列提示信息则表示配置成功。

```
mvn -v
Apache Maven 3.5.2 (138edd61fd100ec658bfa2d307c43b76940a5d7d; 2017-10-18T15:58:13+08:00)
Maven home: E:\apache-maven-3.5.2\bin\..
Java version: 1.8.0_221, vendor: Oracle Corporation
Java home: C:\Java\jdk1.8.0_221\jre
Default locale: zh_CN, platform encoding: GBK
OS name: "windows 10", version: "10.0", arch: "amd64", family: "windows"
```

11.1.4 在 IDEA 中配置 Maven

在 IDEA 中配置 Maven 的步骤如下。

（1）选择 File 菜单→Other Settings 选项→Settings for New Projects 选项，该操作对以后创建的新项目起作用。选择 File 菜单→Settings 选项，该操作对当前项目起作用。配置时要注意，建议两个都配置一下。这里先配置当前项目，选择方式见步骤（2）。

（2）在打开的窗口中，选择 Build，Execution，Deployment 选项→Build Tools 选项→Maven 选项。

指定 Maven 的安装目录、配置文件地址和本地仓库地址，注意要勾选 Override。

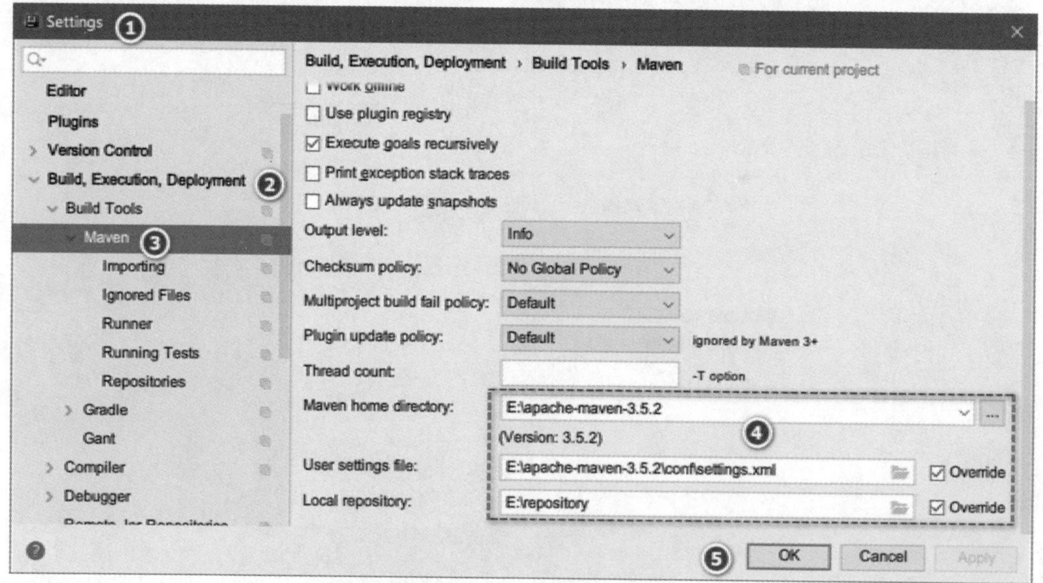

图 11-4　在 IDEA 中配置 Maven

（3）在该窗口中，再选择 Build，Execution，Deployment 选项→Build Tools 选项→Maven 选项→Runner 选项，设置 Maven 启动虚拟机的 VMOption 选项。将该选项设置为所有资源先从本地仓库查找，如果本地仓库中没有才从互联网查找。虚拟机选项的配置如图 11-5 所示，所添加选项如下。

```
-DarchetypeCatalog=internal
```

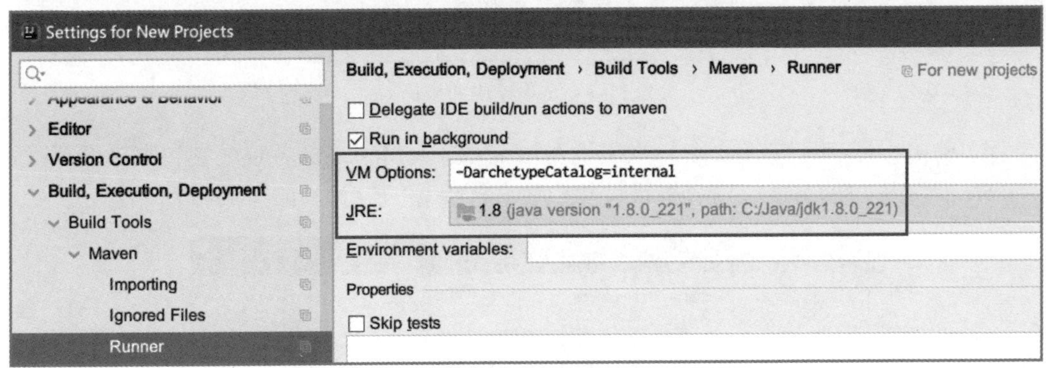

图 11-5　配置虚拟机选项

11.2　使用 Maven 搭建 SSM 环境

视频讲解

11.2.1　创建 Maven 工程

Maven 工程的创建步骤如下。

（1）选择 File 菜单→New 选项→Module 选项。

（2）在打开的窗口中，选择 Maven 选项，勾选 Create from archetype，选中 webapp 模板，如图 11-6 所示。

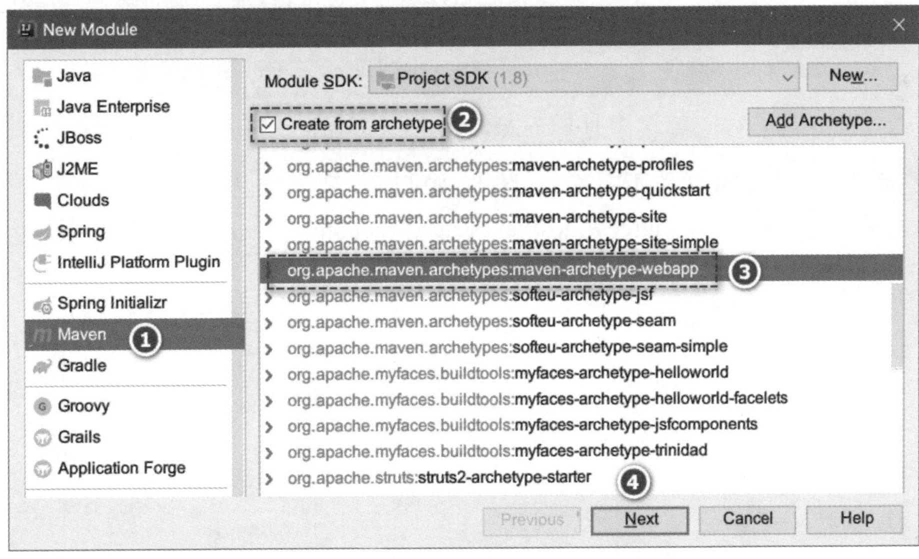

图 11-6　Maven 创建新的 Web 模块

（3）输入当前工程的坐标，这是为了以后放在仓库中给其他工程引用的，代表这个工程在仓库中的唯一坐标。模块坐标设置如图 11-7 所示。

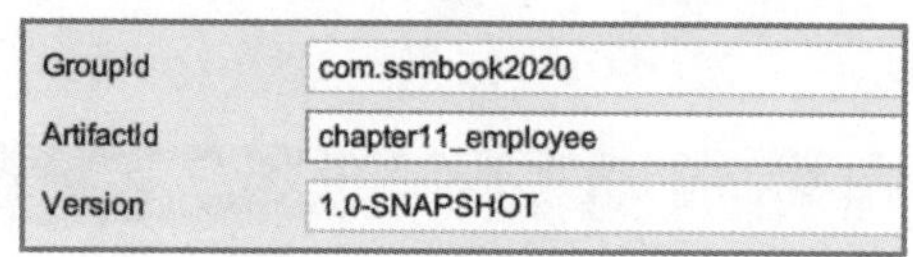

图 11-7　设置模块的坐标

（4）单击“下一步”按钮，会再次出现对话框，确认 Maven 的各项配置参数是否正确。

（5）指定模块的名字、内容根目录和模块文件保存的位置，如图 11-8 所示。

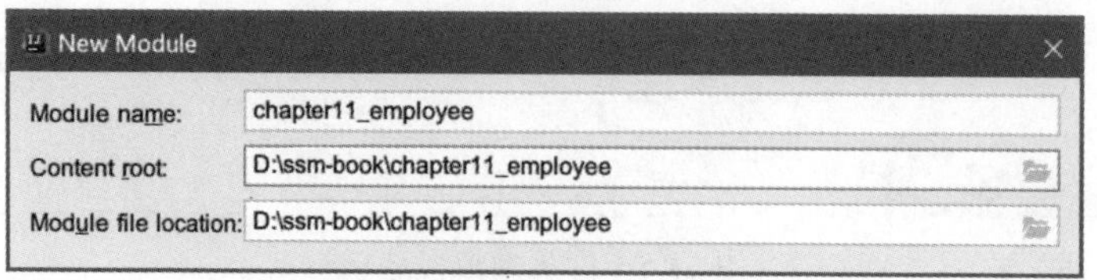

图 11-8　指定模块的名字和位置

11.2.2　完善工程的目录结构

完善工程的目录结构的步骤如下。

（1）一个完整的 Web 目录结构如图 11-9 所示。

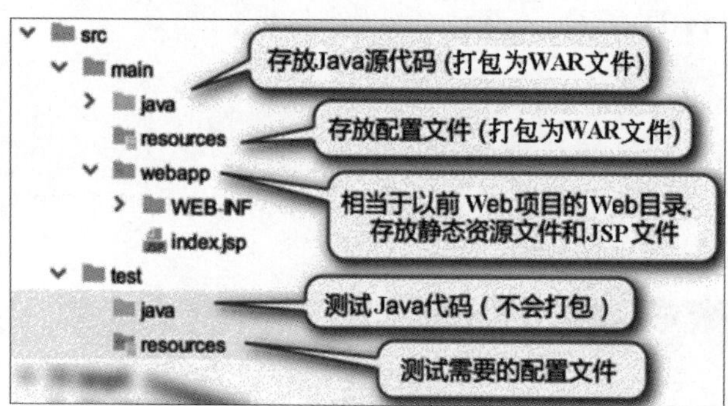

图 11-9　完整的 Maven 目录结构

（2）现在创建的目录结构不完整，如图 11-10 所示，需要自己手动创建其他缺失目录。

（3）手动创建 java、resources、test/java、test/resources 目录，如图 11-11 所示。注意：main 目录与 test 目录平级，都在 src 目录下。

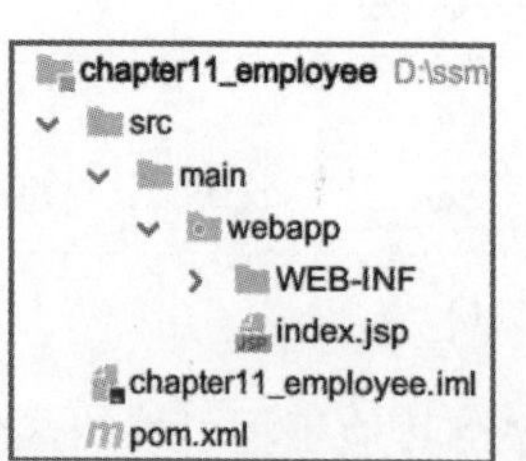

图 11-10　模板创建出来的目录结构

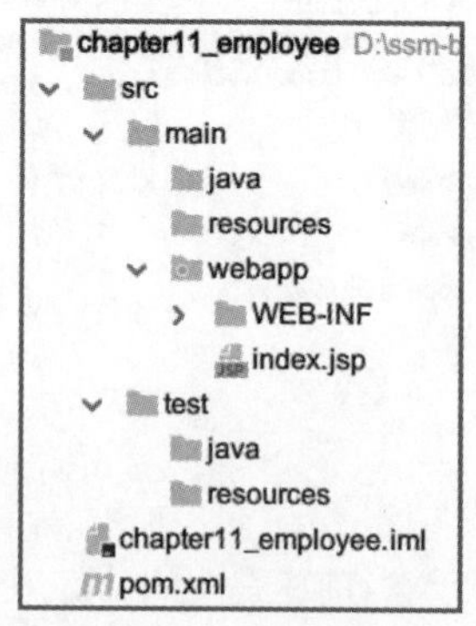

图 11-11　创建新的目录

（4）创建好目录后，点左上角的图标，刷新 Maven 工程，如图 11-12 所示。

（5）左边目录图标的颜色发生改变，说明创建成功了，如图 11-13 所示。

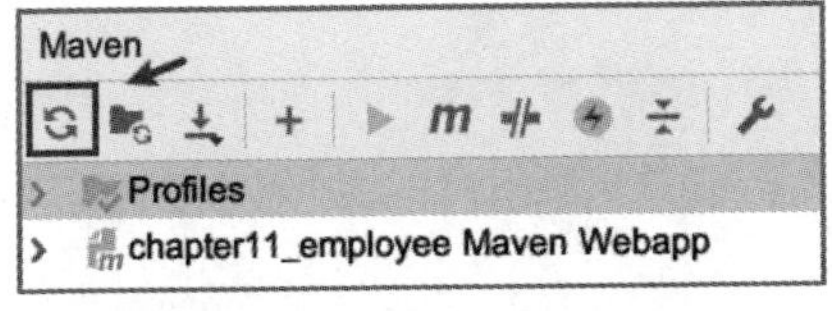

图 11-12　刷新 Maven 工程

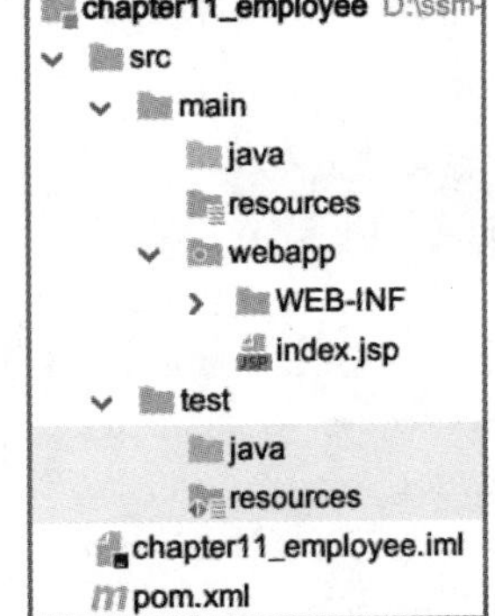

图 11-13　创建成功

（6）如果图标颜色没有变化，就需要在目录上右击选择菜单，手动一个个地标记目录，如图 11-14 所示。

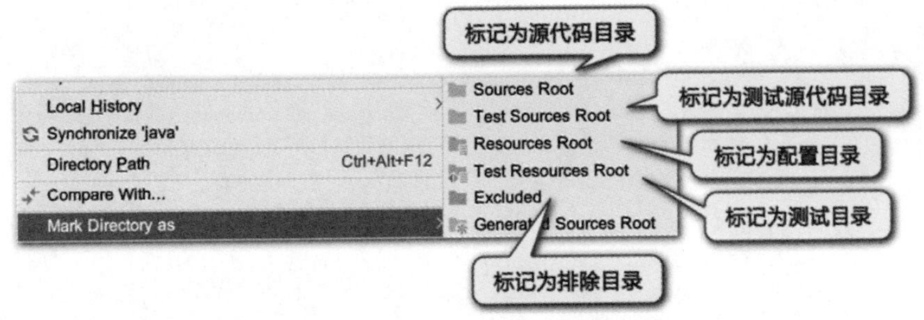

图 11-14　手动标记目录

11.2.3　搭建 SSM 开发环境

创建好 Web 工程以后，就可以开始搭建 SSM 框架开发的工程了，其步骤如下。

（1）打开已经创建好的 pom.xml 文件，删除其中不必要的代码，保留以下内容即可。

```
<?xml version="1.0" encoding="UTF-8"?>
<project xmlns="http://maven.apache.org/POM/4.0.0"
xmlns:xsi="http://www.w3.org/2001/XMLSchema-instance"
        xsi:schemaLocation="http://maven.apache.org/POM/4.0.0
http://maven.apache.org/xsd/maven-4.0.0.xsd">
    <modelVersion>4.0.0</modelVersion>
    <!--当前工程的坐标-->
    <groupId>com.ssmbook2020</groupId>
    <artifactId>chapter11_employee</artifactId>
    <version>1.0-SNAPSHOT</version>
    <!-- 打包的类型: 1. 普通 Java 工程，设置成 JAR 2. Web 工程，设置成 WAR -->
    <packaging>war</packaging>
    <!--当前工程的名字-->
```

```
    <name>chapter11_employee</name>
</project>
```

（2）在 pom.xml 文件中使用 dependencies 元素添加依赖包，每个 dependency 代表一个 JAR 包。

```
<?xml version="1.0" encoding="UTF-8"?>
<project xmlns="http://maven.apache.org/POM/4.0.0"
xmlns:xsi="http://www.w3.org/2001/XMLSchema-instance"
        xsi:schemaLocation="http://maven.apache.org/POM/4.0.0
http://maven.apache.org/xsd/maven-4.0.0.xsd">
    <modelVersion>4.0.0</modelVersion>
    <!--当前工程的坐标-->
    <groupId>com.ssmbook2020</groupId>
    <artifactId>chapter11_employee</artifactId>
    <version>1.0-SNAPSHOT</version>
    <!-- 打包的类型：1. 普通 Java 工程，设置成 JAR 2. Web 工程，设置成 WAR -->
    <packaging>war</packaging>
    <!--当前工程的名字-->
    <name>chapter11_employee</name>
    <!-- 当前项目中依赖的所有 JAR 包 -->
    <dependencies>
        <!--SpringMVC 的包-->
        <dependency>
            <groupId>org.springframework</groupId>
            <artifactId>spring-webmvc</artifactId>
            <version>5.2.0.RELEASE</version>
        </dependency>
        <!--Servlet 包-->
        <dependency>
            <groupId>javax.servlet</groupId>
            <artifactId>javax.servlet-api</artifactId>
            <version>3.1.0</version>
            <scope>provided</scope>
        </dependency>
        <!--JSON 包-->
        <dependency>
            <groupId>com.fasterxml.jackson.core</groupId>
            <artifactId>jackson-databind</artifactId>
            <version>2.9.5</version>
        </dependency>
        <!--Druid 连接池-->
        <dependency>
            <groupId>com.alibaba</groupId>
            <artifactId>druid</artifactId>
            <version>1.1.12</version>
        </dependency>
```

```
<!--MySQL 驱动，如果使用的是 8.0 版，则需要更换驱动-->
<dependency>
    <groupId>mysql</groupId>
    <artifactId>mysql-connector-java</artifactId>
    <version>5.1.47</version>
</dependency>
<!--访问 JDBC 的包-->
<dependency>
    <groupId>org.springframework</groupId>
    <artifactId>spring-jdbc</artifactId>
    <version>5.2.0.RELEASE</version>
</dependency>
<!-- 事务处理包 -->
<dependency>
    <groupId>org.springframework</groupId>
    <artifactId>spring-tx</artifactId>
    <version>5.2.0.RELEASE</version>
</dependency>
<!-- Mybatis 整合 Spring 的包 -->
<dependency>
    <groupId>org.mybatis</groupId>
    <artifactId>mybatis-spring</artifactId>
    <version>1.3.1</version>
</dependency>
<!-- Lombok，可以让实体类省略 set 和 get 方法 -->
<dependency>
    <groupId>org.projectlombok</groupId>
    <artifactId>lombok</artifactId>
    <version>1.18.12</version>
    <scope>provided</scope>
</dependency>
<!-- Mybatis 框架 -->
<dependency>
    <groupId>org.mybatis</groupId>
    <artifactId>mybatis</artifactId>
    <version>3.5.1</version>
</dependency>
<!--JUnit 测试-->
<dependency>
    <groupId>junit</groupId>
    <artifactId>junit</artifactId>
    <version>4.12</version>
    <scope>test</scope>
</dependency>
<!--Spring 整合 JUnit-->
<dependency>
    <groupId>org.springframework</groupId>
```

```
            <artifactId>spring-test</artifactId>
            <version>5.2.0.RELEASE</version>
            <scope>test</scope>
        </dependency>
    </dependencies>
    <build>
        <plugins>
            <!-- Java 编译插件 -->
            <plugin>
                <groupId>org.apache.maven.plugins</groupId>
                <artifactId>maven-compiler-plugin</artifactId>
                <version>3.8.0</version>
                <configuration>
                    <source>1.8</source>
                    <target>1.8</target>
                    <encoding>UTF-8</encoding>
                </configuration>
            </plugin>
            <!--Tomcat7 插件-->
            <plugin>
                <groupId>org.apache.tomcat.maven</groupId>
                <artifactId>tomcat7-maven-plugin</artifactId>
                <version>2.2</version>
                <configuration>
                    <port>8080</port>
                    <path>/</path>
                    <uriEncoding>UTF-8</uriEncoding>
                    <server>tomcat7</server>
                </configuration>
            </plugin>
        </plugins>
    </build>
</project>
```

（3）有时需要点一下右上角的 Reimport 按钮刷新一次，如图 11-15 所示。

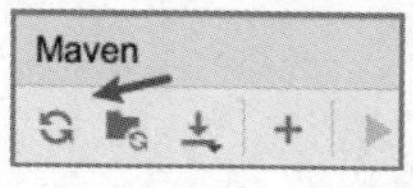

图 11-15　刷新 SSM 工程

（4）查看是否所有的依赖 JAR 包都出现了，这里可能需要等待下载 JAR 包。如果是第一次运行，可能下载的时间比较长，因为此时所有的 JAR 包都要从远程仓库下载，以后就不需要了。SSM 的 JAR 包如图 11-16 所示。

（5）在 main/resources 上右击，选择 New 选项→XML Configuration File 选项→Spring Config 选项，创建 springMVC.xml，这是 SpringMVC 的配置文件。在 springMVC.xml 中需要导入 context 和 mvc 命名空间，按 Alt+Enter 键会自动导入，注意不要导错为同名的命名

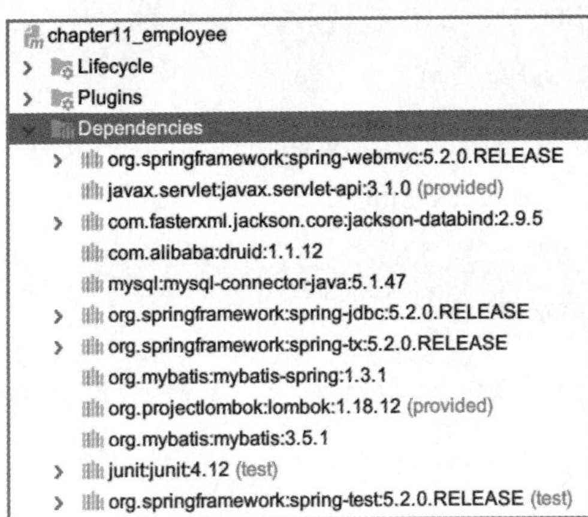

图 11-16　SSM 的 JAR 包

空间。这时不用理会 com.ssmbook2020.controller 处的报红，因为还没有创建这个包。

```
    <?xml version="1.0" encoding="UTF-8"?>
    <beans xmlns="http://www.springframework.org/schema/beans"
          xmlns:xsi="http://www.w3.org/2001/XMLSchema-instance"
    xmlns:context="http://www.springframework.org/schema/context"
          xmlns:mvc="http://www.springframework.org/schema/mvc"
          xsi:schemaLocation="http://www.springframework.org/schema/beans
http://www.springframework.org/schema/beans/spring-beans.xsd
http://www.springframework.org/schema/context
https://www.springframework.org/schema/context/spring-context.xsd
http://www.springframework.org/schema/mvc
https://www.springframework.org/schema/mvc/spring-mvc.xsd">
    <!-- 1. 控制器包的注解扫描 -->
    <context:component-scan base-package="com.ssmbook2020.controller"/>
    <!-- 2. 视图解析器 -->
    <bean class="org.springframework.web.servlet.view.
InternalResourceViewResolver"/>
    <!-- 3. 注册 mvc 驱动 -->
    <mvc:annotation-driven/>
    <!-- 4. 静态资源由默认 Servlet 处理 -->
    <mvc:default-servlet-handler/>
    </beans>
```

（6）在 main/resources 上右击，选择 New 选项→File 选项，创建 jdbc.properties 文件，这是数据库的配置文件。本书用的是 5.x 的 MySQL 数据库版本。

```
    # 数据库的驱动名，如果是 8.0 版可能有所区别
    jdbc.driver=com.mysql.jdbc.Driver
    # 连接字符串，如果是 8.0 版请参考前面的章节，本章数据库名为 db_emp
    jdbc.url=jdbc:mysql://localhost:3306/db_emp
    # 用户名
    jdbc.username=root
```

```
# 密码
jdbc.password=root
```

（7）在 main/resources 上右击，创建 MyBatis 的核心配置文件 mybatis-config.xml，会发现大部分配置都在 applicationContext.xml 中。

```
<?xml version="1.0" encoding="UTF-8" ?>
<!DOCTYPE configuration
       PUBLIC "-//mybatis.org//DTD Config 3.0//EN"
       "http://mybatis.org/dtd/mybatis-3-config.dtd">
<configuration>
   <settings>
       <!--在控制台显示 SQL 语句-->
       <setting name="logImpl" value="STDOUT_LOGGING"/>
       <!--映射下画线为驼峰命名法-->
       <setting name="mapUnderscoreToCamelCase" value="true"/>
   </settings>
   <!—定义实体类别名-->
   <typeAliases>
       <package name="com.ssmbook2020.entity"/>
   </typeAliases>
</configuration>
```

（8）在 main/resources 上右击，选择 New 选项→XML Configuration File 选项→Spring Config 选项，创建 applicationContext.xml 文件，这是 Spring 的配置文件。注意要导入 context 和 tx 命名空间，且不能导入错误。如果 com.ssmbook2020.service 处报红，同样不用理会。

```
<?xml version="1.0" encoding="UTF-8"?>
<beans xmlns="http://www.springframework.org/schema/beans"
      xmlns:xsi="http://www.w3.org/2001/XMLSchema-instance"
xmlns:context="http://www.springframework.org/schema/context"
      xmlns:tx="http://www.springframework.org/schema/tx"
      xsi:schemaLocation="http://www.springframework.org/schema/beans
http://www.springframework.org/schema/beans/spring-beans.xsd
http://www.springframework.org/schema/context
https://www.springframework.org/schema/context/spring-context.xsd
http://www.springframework.org/schema/tx
http://www.springframework.org/schema/tx/spring-tx.xsd">
   <!-- 扫描业务类 -->
   <context:component-scan base-package="com.ssmbook2020.service"/>
   <!-- 加载属性配置文件，并且将键和值放在容器中 -->
   <context:property-placeholder location="classpath:jdbc.properties"/>
   <!-- 配置数据源对象 -->
   <bean class="com.alibaba.druid.pool.DruidDataSource" id="dataSource">
       <!--连接字符串 -->
       <property name="url" value="${jdbc.url}"/>
       <!--用户名-->
```

```
            <property name="username" value="${jdbc.username}"/>
            <!--密码-->
            <property name="password" value="${jdbc.password}"/>
            <!--驱动名字-->
            <property name="driverClassName" value="${jdbc.driver}"/>
        </bean>
        <!-- 会话工厂类 -->
        <bean class="org.mybatis.spring.SqlSessionFactoryBean">
            <!--指定数据源-->
            <property name="dataSource" ref="dataSource"/>
            <!--指定 mybatis 的核心配置文件-->
            <property name="configLocation" value="classpath:mybatis-config.
xml"/>
        </bean>
        <!-- DAO 代理类 -->
        <bean class="org.mybatis.spring.mapper.MapperScannerConfigurer">
            <!-- 对 dao 包中所有的接口生成代理对象 -->
            <property name="basePackage" value="com.ssmbook2020.dao"/>
        </bean>
        <!-- 事务管理器 -->
        <bean class="org.springframework.jdbc.datasource.
DataSourceTransactionManager" id="transactionManager">
            <property name="dataSource" ref="dataSource"/>
        </bean>
        <!-- 注解式事务 -->
        <tx:annotation-driven/>
    </beans>
```

（9）在 webapp/WEB-INF/web.xml 中编写配置文件，整合 Spring 和 Spring MVC 框架。

```
    <?xml version="1.0" encoding="UTF-8"?>
    <web-app xmlns:xsi="http://www.w3.org/2001/XMLSchema-instance"
            xmlns="http://java.sun.com/xml/ns/javaee"
            xsi:schemaLocation="http://java.sun.com/xml/ns/javaee
http://java.sun.com/xml/ns/javaee/web-app_2_5.xsd"
            version="2.5">
        <!--Spring MVC 的 Servlet 配置-->
        <servlet>
            <servlet-name>dispatcherServlet</servlet-name>

            <servlet-class>org.springframework.web.servlet.DispatcherServlet
</servlet-class>
            <init-param>
                <param-name>contextConfigLocation</param-name>
                <param-value>classpath:springMVC.xml</param-value>
            </init-param>
        </servlet>
        <servlet-mapping>
```

```
        <servlet-name>dispatcherServlet</servlet-name>
        <url-pattern>/</url-pattern>
    </servlet-mapping>
    <!-- Spring 的配置-->
    <listener>
  <listener-class>org.springframework.web.context.ContextLoaderListener
</listener-class>
    </listener>
    <context-param>
        <param-name>contextConfigLocation</param-name>
        <param-value>classpath:applicationContext.xml</param-value>
    </context-param>
  </web-app>
```

（10）在 src/main/java 下创建 com.ssmbook2020.controller、com.ssmbook2020.dao 和 com.ssmbook2020.service 包，此时的目录结构如图 11-17 所示。至此，SSM 开发环境就搭建好了。

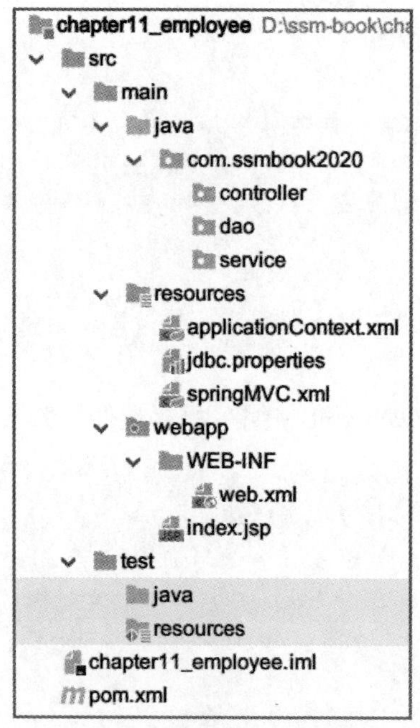

图 11-17　搭建好的目录结构

11.2.4　在 Tomcat 中部署运行

因为在 pom.xml 中配置了 Tomcat7 的插件，直接选择 Plugins 选项→tomcat7 选项→tomcat7:run 选项运行即可，如图 11-18 所示。

单击控制台下面出现的链接，打开浏览器 http://localhost:8080/，在浏览器中看到如图 11-19 所示的页面，这是 Tomcat 默认打开了 webapp 下的 index.jsp 文件。

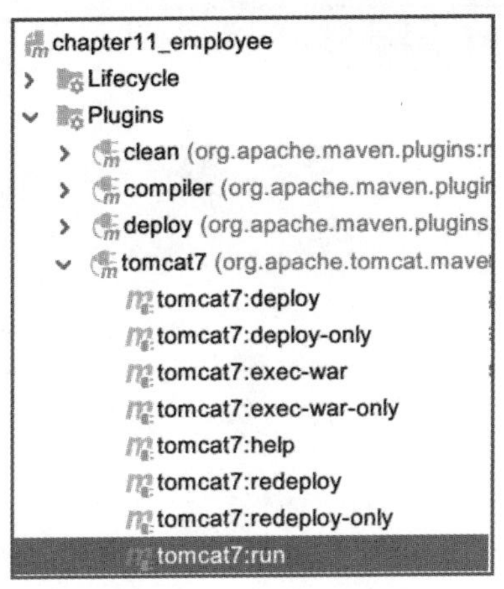

图 11-18　使用插件运行

图 11-19　首页效果

11.3　员工管理系统的实现

11.3.1　项目需求

（1）开发员工管理模块，实现员工的增、删、改、查的基本操作。

（2）实现员工多条件组合查询的复杂操作。

11.3.2　运行效果

（1）查询所有的员工，有分页功能，如图 11-20 所示。

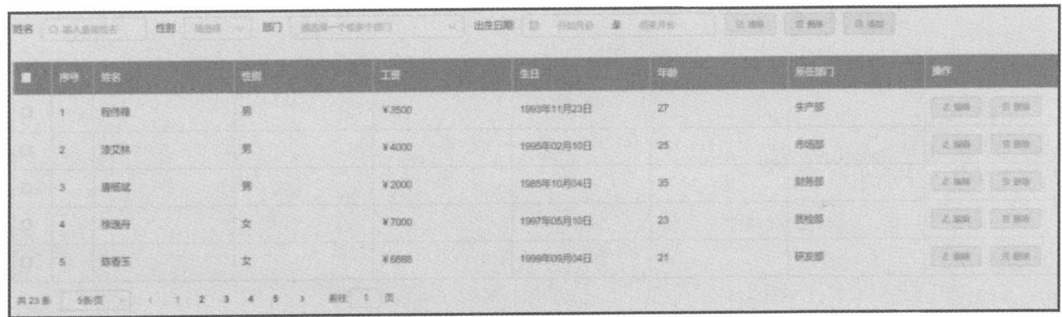

图 11-20　查询所有员工

（2）支持多条件模糊查询，如图 11-21 所示。

图 11-21　多条件查询

（3）添加和编辑员工信息，有完整的校验，如图 11-22 所示。

（4）在编辑页面加载当前员工的信息，将其显示在该窗口中，如图 11-23 所示。

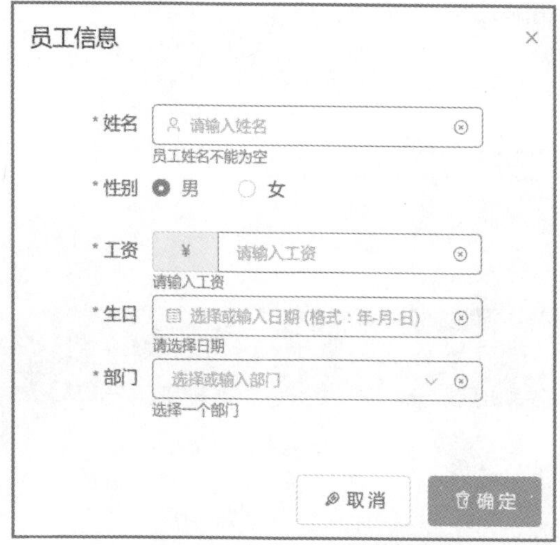

图 11-22　添加和编辑页面

图 11-23　编辑页面加载员工信息

（5）有批量删除的功能，如图 11-24 所示。

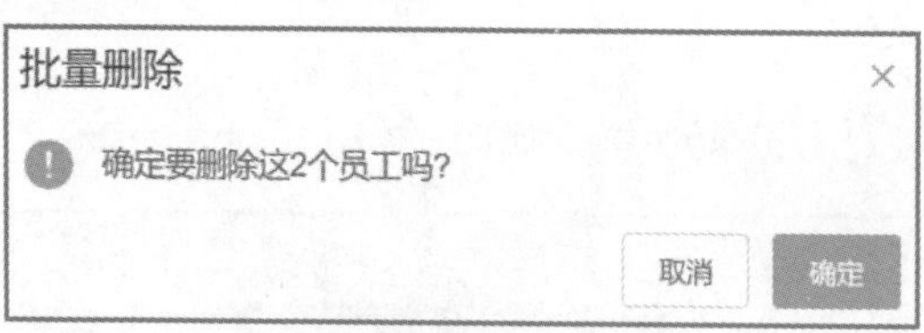

图 11-24　删除多个员工

11.3.3　数据库设计

创建名为 db_emp 的数据库。

```
create database db_emp;
```

执行以下 SQL 语句代码。

```
-- 部门表
CREATE TABLE depart (
```

```
  id   INT PRIMARY KEY AUTO_INCREMENT,
  'name' VARCHAR(20) NOT NULL
);
-- 员工表
CREATE TABLE employee (
  id   INT PRIMARY KEY AUTO_INCREMENT,
  'name'     VARCHAR(20),
  sex     CHAR(1) DEFAULT '男',
  salary   DOUBLE,  -- 工资
  birthday  DATE,  -- 生日
  depart_id INT,
  FOREIGN KEY (depart_id) REFERENCES depart(id)
);
-- 添加部门数据
INSERT INTO depart ('name') VALUES ('公关部'),('生产部'),('质检部'),('财务部'),
('市场部'),('销售部'),('研发部'),('行政部');
SELECT * FROM depart;
-- 添加员工数据
INSERT INTO EMPLOYEE ('name', sex, salary, birthday, depart_id) VALUES
('程伟锋', '男', 3500, '1993-11-23', 2),('漆艾林', '男', 4000, '1995-02-10', 5),
('唐杨斌' , '男', 2000, '1985-10-4',4),('徐逸舟' ,'女', 7000, '1997-05-10',3),
('陈香玉','女', 6888, '1999-09-04',7),('邹丽娜','女', 12000, '1988-04-24',1),
('蒋思思' ,'女', 6000, '1987-10-3',8),('陈锡猛' ,'男', 2700, '2000-11-12',3),
('曾刚' ,'男', 3600, '1994-05-20', 2),('罗淑丽' ,'女', 10000, '1976-04-24',1),
('牛佳' ,'女', 6600, '1995-11-3',8),('何吉鹏' ,'男', 4600, '2001-11-12',3),
('陈浩俊' ,'男', 7800, '1997-08-20', 2),('刘宇轩','男',9000,'1990-05-08',6),
('何耀辉' ,'男', 16000, '1998-04-24',1),('刘语芯' ,'女', 6700, '1997-07-3',8),
('张美美','女', 8700, '2002-10-12',3),('胡兰兰' ,'女', 9600, '1996-12-20', 2),
('钟美美','女',8000,'1982-10-25',3),('程伟锋','男',35000,'1993-11-11',2),
('马小洁','女',4500,'1983-06-23',5),('猪八戒','男',3333,'2000-09-29',2),
('刘二小','女',15000,'1998-11-12',4);
-- 查询所有员工
SELECT * FROM employee;
```

表之间的关系如图 11-25 所示。

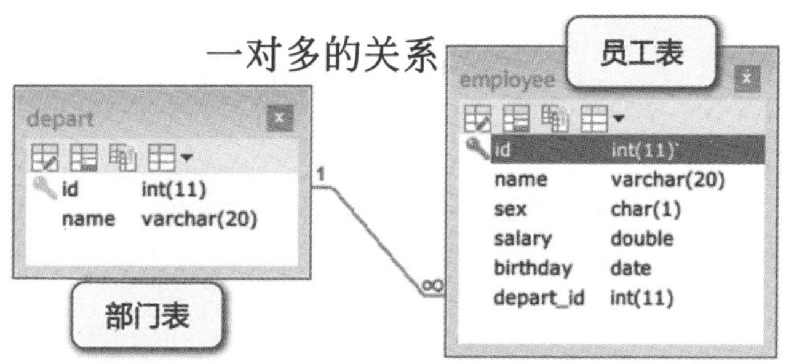

图 11-25　表之间的关系

11.3.4 Lombok 插件

1. Lombok 介绍

在以前的 Java 项目中，充斥着太多不友好的代码，如 POJO 的 getter/setter/toString、异常处理代码、I/O 流的关闭操作等。这些样板代码既没有技术含量，又影响着代码的美观，于是 Lombok 应运而生。Lombok 是一个插件，用途是使用注解给类里面的字段自动地加上属性、构造器、toString 方法和 Equals 方法等。如果字段发生更改，Lombok 会立即同步改变，以保持代码的一致性。

2. Lombok 使用

（1）在 pom.xml 中导入 Lombok 的坐标，实际在员工管理系统中已经导入了。

```
<dependency>
    <groupId>org.projectlombok</groupId>
    <artifactId>lombok</artifactId>
    <version>1.18.12</version>
    <scope>provided</scope>
</dependency>
```

（2）在 IDEA 中安装 Lombok 插件，选择 File 选项→Settings 选项→Plugins 选项，安装完成后要重启一次 IDEA，如图 11-26 所示。

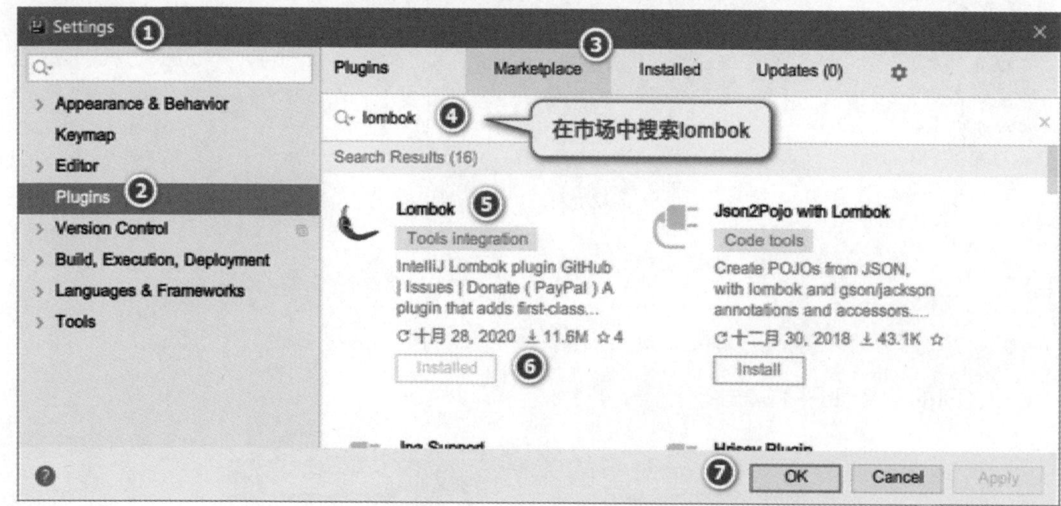

图 11-26 安装 Lombok 插件

（3）在 IDEA 中配置 Lombok 支持，选择 File 选项→Settings 选项→Annocation Processors 选项，开启 Annocation Processors，如图 11-27 所示。开启该项是为了让 Lombok 注解在编译阶段起到作用。

3. Lombok 常用注解

Lombok 的常用注解如表 11-3 所示。

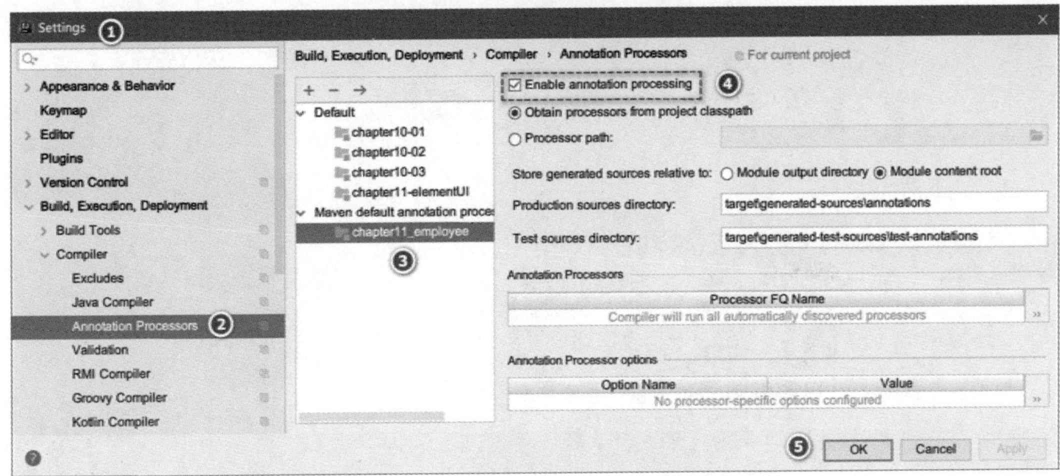

图 11-27 配置开启 Annocation Processors

表 11-3 Lombok 的常用注解

注 解	作 用
@Getter/@Setter	放在成员变量上：给这个成员变量生成 get 和 set 方法。 放在类上：给这个类中所有的成员变量生成 get 和 set 方法
@ToString	放在类上：将类中所有的属性生成 toString 方法。 of 属性：指定只包含哪些属性。 exclude 属性：排除哪些属性
@EqualsAndHashCode	放在类上：生成 equals 和 hashCode 方法
@NoArgsConstructor	放在类上：生成无参的构造方法
@AllArgsConstructor	放在类上：生成全部参数的构造方法
@Data	放在类上：会为类的所有属性自动生成 setter/getter、equals、canEqual、hashCode、toString 方法。如果属性是 final 类型，则不会生成

11.3.5 实体类对象

视频讲解

创建实体类，因为有两张表则对应两个实体类。实体类的包如图 11-28 所示。

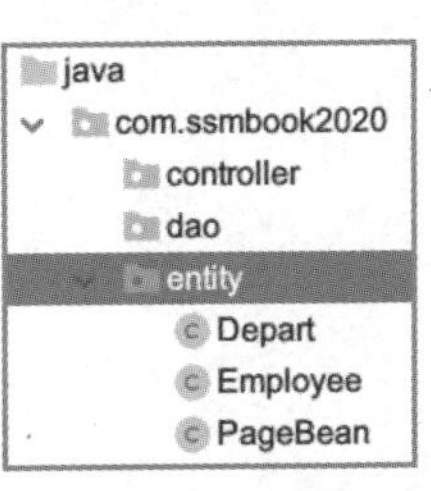

图 11-28 实体类的包

1. 部门对象

```
package com.ssmbook2020.entity;
import lombok.Data;
@Data //自动生成 getter 和 setter 方法
```

```
public class Depart {
    private Integer id;      //编号
    private String name;     //姓名
}
```

2. 员工对象

```
package com.ssmbook2020.entity;
import com.fasterxml.jackson.annotation.JsonFormat;
import lombok.Data;
//日期使用 sql 的 Date 对象
import java.sql.Date;
@Data
public class Employee {
    private Integer id;           //编号
    private String name;          //姓名
    private String sex;           //性别
    private Double salary;        //薪水
    //用于 JSON 转换时指定日期格式化，这里要指定时区，不然转换出来的日期会少一天
    @JsonFormat(pattern = "yyyy 年 mm 月 dd 日", timezone = "GMT+8")
    private Date birthday;
    private Integer departId;     //对应部门的外键
    private Depart depart;        //对应部门对象
    //年龄字段在数据库中没有，通过生日计算得出
    private Integer age;
}
```

3. 分页对象

因为用到了分页，这里抽象出一个分页的对象，其名为 PageBean。MyBatis 框架中有些插件也自带类似的分页类对象。

```
package com.ssmbook2020.entity;
import lombok.Data;
import java.util.List;
/**
 * 分页对象：封装一页所有数据
 */
@Data
public class PageBean<T>  {
    /*
    一共是 8 个属性，分成以下三大类。
     (1) 从数据库中查询出来的属性：data、count。
     (2) 由用户从浏览器提交过来的属性：current、size。
     (3) 由其他的属性计算出来的属性：first、 previous、next、total，写在 get 方法中
     */
    private List<T> data;    //封装一页的数据
    private int count;       //总记录数
```

```
    private int current;    //当前第几页
    private int size;       //每页的大小
    private int first;      //第一页
    private int previous;   //上一页
    private int next;       //下一页
    private int total;      //总页数/最后一页
    //构造方法传入 4 个属性
    public PageBean(List<T> data, int count, int current, int size) {
        this.data = data;
        this.count = count;
        this.current = current;
        this.size = size;
    }
    public PageBean() {
    }
    /**
     * 获取第一页
     */
    public int getFirst() {
        return 1;
    }
    /**
     * 获取上一页：如果当前页大于 1，上一页就等于当前页减一，否则返回第一页
     */
    public int getPrevious() {
        if (getCurrent() > 1) {
            return getCurrent() - 1;
        } else {
            return 1;
        }
    }
    /**
     * 计算下一页：如果当前页小于最后一页，下一页就等于当前页加一，否则返回最后一页
     */
    public int getNext() {
        if (getCurrent() < getTotal()) {
            return getCurrent() + 1;
        } else {
            return getTotal();
        }
    }
    /**
     * 计算总页数：如果总记录数能够整除页大小，则为该页数。如果不能整除，页数就加一
     */
    public int getTotal() {
        if (getCount() % getSize() == 0) {
            return getCount() / getSize();
```

```
        } else {
            return getCount() / getSize() + 1;
        }
    }
    //注意：输出的是get方法的值才是准确的，因为在get中计算了属性值，没有给成员变量赋值
    @Override
    public String toString() {
        return "PageBean{" +
                "data=" + getData() +
                ", count=" + getCount() +
                ", current=" + getCurrent() +
                ", size=" + getSize() +
                ", first=" + getFirst() +
                ", previous=" + getPrevious() +
                ", next=" + getNext() +
                ", total=" + getTotal() +
                '}';
    }
}
```

11.3.6 数据访问层

1. 员工类的 DAO 接口

```
package com.ssmbook2020.dao;
import com.ssmbook2020.entity.Employee;
import org.apache.ibatis.annotations.Delete;
import org.apache.ibatis.annotations.Insert;
import org.apache.ibatis.annotations.Param;
import org.apache.ibatis.annotations.Update;
import java.util.List;
import java.util.Map;
/**
 * 员工的DAO类
 */
public interface EmployeeDao {
    /**
     * 查询一页员工和部门信息
     */
    List<Employee> findEmployees(@Param("param") Map<String, Object> param);

    /**
     * 查询总记录数
     */
    int findCount(@Param("param") Map<String, Object> param);

    /**
```

```
     * 插入员工
     */
    @Insert("INSERT INTO employee VALUES (NULL,#{name},#{sex},#{salary},
#{birthday},#{departId})")
    int save(Employee employee);

    /**
     * 更新员工
     */
    @Update("UPDATE employee SET name = #{name},sex = #{sex},salary =
#{salary},birthday = #{birthday},depart_id = #{departId} WHERE id = #{id}")
    int update(Employee employee);

    /**
     * 删除员工
     */
    @Delete("delete from employee where id=#{id}")
    int delete(Integer id);

    /**
     * 删除多个员工
     */
    int deleteEmployees(Integer[] employeeIds);
}
```

2. 部门的 DAO 接口

```
package com.ssmbook2020.dao;
import com.ssmbook2020.entity.Depart;
import org.apache.ibatis.annotations.*;
import java.util.List;
/**
 * 部门的 DAO 类
 */
public interface DepartDao {
    /**
     * 查询所有部门
     */
    @Select("select * from depart")
    List<Depart> findAllDeparts();
}
```

3. 员工类的映射配置文件

员工类使用了条件组合查询，需要动态生成 SQL 语句，这里使用了 XML 配置文件。具体操作步骤如下。

（1）在 resources 目录下右击，选择 New 选项→Directory 选项，创建目录，如图 11-29

所示。注意：这里目录之间使用/分隔，而不是包名的点号。

（2）创建出来的文件夹看起来像一个包的结构，如图 11-30 所示。

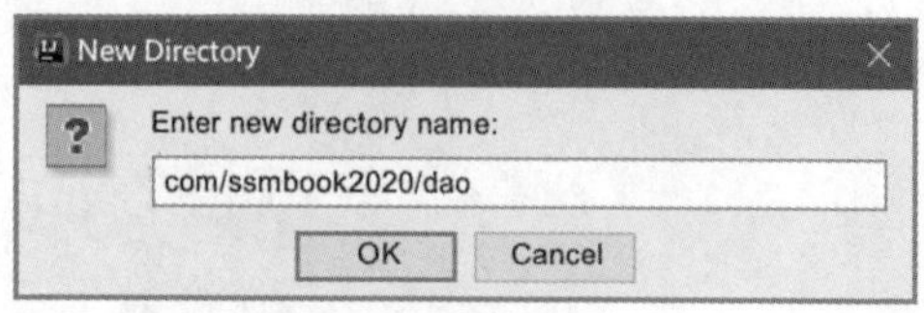

图 11-29　创建目录结构

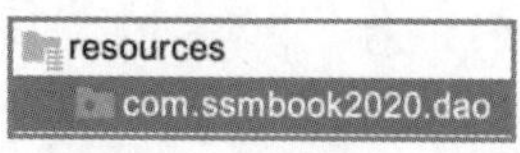

图 11-30　创建后图标

（3）创建 XML 实体映射文件 EmployeeDao.xml，内容如下。

```
<?xml version="1.0" encoding="UTF-8" ?>
<!DOCTYPE mapper
        PUBLIC "-//mybatis.org//DTD Mapper 3.0//EN"
        "http://mybatis.org/dtd/mybatis-3-mapper.dtd">
<mapper namespace="com.ssmbook2020.dao.EmployeeDao">
    <!--员工的映射-->
    <resultMap id="employeeMap" type="employee">
        <result column="did" property="depart.id"/>
        <result column="dname" property="depart.name"/>
    </resultMap>

    <!--查询条件的代码段-->
    <sql id="paramSql">
        <where>
            <!-- 这里要指定是哪个表的 name -->
            <if test="param.name!=null and param.name!=''">
                e.name like "%"#{param.name}"%"
            </if>
            <if test="param.sex!=null and param.sex!=''">
                and sex = #{param.sex}
            </if>
            <!-- 可以查询多个部门 -->
            <if test="param.departId!=null and param.departId.size>0">
                and depart_id in
                <foreach collection="param.departId" open="(" item="id"
separator="," close=")">
                    #{id}
                </foreach>
            </if>
            <if test="param.birthday!=null and param.birthday.size > 0">
                <!-- 这里生日是一个 List 集合，第一个日期是 0，第二个日期是 1 -->
                and birthday between #{param.birthday[0]} and #{param.birthday[1]}
            </if>
        </where>
    </sql>
```

```
    <!--删除多个员工-->
    <delete id="deleteEmployees">
       delete from employee where id in
       <foreach collection="array" open="(" separator="," item="id" close=")">
          #{id}
       </foreach>
    </delete>

    <!--查询一页的员工数据，按员工 id 排序，年龄通过生日计算出来 -->
    <select id="findEmployees" resultMap="employeeMap">
       SELECT e.*, TIMESTAMPDIFF(YEAR,birthday,NOW()) AS age, d.id did,
d.name dname
       FROM employee e LEFT JOIN depart d ON e.depart_id = d.id
       <include refid="paramSql"/>
       ORDER BY e.id LIMIT #{param.begin},#{param.size}
    </select>

    <!--查询总记录数-->
    <select id="findCount" resultType="java.lang.Integer">
       select count(*) from employee e
       <include refid="paramSql"/>
    </select>
 </mapper>
```

11.3.7 业务层

1. 员工的业务接口

```
package com.ssmbook2020.service;

import com.ssmbook2020.entity.Employee;
import com.ssmbook2020.entity.PageBean;
import java.util.Map;

public interface EmployeeService {
   /**
    * 查询一页数据，封装成一个页面对象
    * @param param 查询条件
    */
   PageBean<Employee> findPageBean(Map<String, Object> param);
   /**
    * 添加员工
    */
   int save(Employee employee);
   /**
    * 更新员工
    */
```

```
    int update(Employee employee);
    /**
     * 删除员工
     */
    int delete(Integer id);
    /**
     * 删除多个员工
     */
    int deleteEmployees(Integer[] employeeIds);
}
```

2. 员工的业务实现类

```
package com.ssmbook2020.service.impl;
import com.ssmbook2020.dao.EmployeeDao;
import com.ssmbook2020.entity.Employee;
import com.ssmbook2020.entity.PageBean;
import com.ssmbook2020.service.EmployeeService;
import org.springframework.beans.factory.annotation.Autowired;
import org.springframework.stereotype.Service;
import java.util.List;
import java.util.Map;

@Service
public class EmployeeServiceImpl implements EmployeeService {

    @Autowired
    private EmployeeDao employeeDao;

    @Override
    public PageBean<Employee> findPageBean(Map<String, Object> param) {
        //获取 current 和 size 的值
        int current = (int) param.get("current");
        int size = (int) param.get("size");
        int begin = (current - 1) * size;
        //计算查询的起始行
        param.put("begin", begin);
        List<Employee> data = employeeDao.findEmployees(param);
        int count = employeeDao.findCount(param);
        return new PageBean<>(data, count, current, size);
    }

    /**
     * 添加员工
     */
    @Override
    public int save(Employee employee) {
```

```
        return employeeDao.save(employee);
    }

    /**
     * 更新员工
     */
    @Override
    public int update(Employee employee) {
        return employeeDao.update(employee);
    }

    /**
     * 删除员工
     */
    @Override
    public int delete(Integer id) {
        return employeeDao.delete(id);
    }

    /**
     * 删除多个员工
     */
    @Override
    public int deleteEmployees(Integer[] employeeIds) {
        return employeeDao.deleteEmployees(employeeIds);
    }
}
```

3. 部门的业务接口

```
package com.ssmbook2020.service;
import com.ssmbook2020.entity.Depart;
import java.util.List;
/**
 * 部门的业务层
 */
public interface DepartService {
    /**
     * 查询所有部门
     */
    List<Depart> findAllDeparts();
}
```

4. 部门的业务实现类

```
package com.ssmbook2020.service.impl;

import com.ssmbook2020.dao.DepartDao;
```

```
import com.ssmbook2020.entity.Depart;
import com.ssmbook2020.service.DepartService;
import org.springframework.beans.factory.annotation.Autowired;
import org.springframework.stereotype.Service;
import java.util.List;

@Service
public class DepartServiceImpl implements DepartService {
    @Autowired
    private DepartDao departDao;
    /**
     * 查询所有部门
     */
    @Override
    public List<Depart> findAllDeparts() {
        return departDao.findAllDeparts();
    }
}
```

视频讲解

11.3.8 测试业务层

1. 测试类

在 test/java 目录下创建测试类 com.ssmbook2020.test.TestEmployee.java，代码如下。

```
package com.ssmbook2020.test;

import com.ssmbook2020.entity.Employee;
import com.ssmbook2020.entity.PageBean;
import com.ssmbook2020.service.EmployeeService;
import org.junit.Test;
import org.junit.runner.RunWith;
import org.springframework.beans.factory.annotation.Autowired;
import org.springframework.test.context.ContextConfiguration;
import org.springframework.test.context.junit4.SpringJUnit4ClassRunner;
import java.util.HashMap;
import java.util.List;

@RunWith(SpringJUnit4ClassRunner.class) //指定 Spring 的运行器
@ContextConfiguration("classpath:applicationContext.xml")  //指定 Spring 的
配置文件
public class TestEmployee {
    @Autowired  //注入业务对象
    private EmployeeService employeeService;

    @Test
    public void testFindPage() {
```

```
        //创建查询条件
        HashMap<String, Object> param = new HashMap<>();
        //查询第一页
        param.put("current", 1);
        //每页显示 3 条
        param.put("size", 3);
        //模糊查询姓名
        param.put("name", "刘");
        //调用业务方法
        PageBean<Employee> pageBean = employeeService.findPageBean(param);
        System.out.println(pageBean);
    }
}
```

2. 控制台输出结果

（1）在控制台可以看到生成的 SQL 语句。

（2）可以看到 PageBean 的结果。

```
    PageBean{data=[Employee(id=14, name=刘宇轩, sex=男, salary=9000.0,
birthday=1990-05-08, departId=6, depart=Depart(id=6, name=销售部), age=30),
Employee(id=16, name=刘语芯, sex=女, salary=6700.0, birthday=1997-07-03,
departId=8, depart=Depart(id=8, name=行政部), age=23), Employee(id=23, name=
刘二小, sex=女, salary=15000.0, birthday=1998-11-12, departId=4,
depart=Depart (id=4, name=财务部), age=22)], count=3, current=1, size=3, first=1,
previous=1, next=1, total=1}
```

11.3.9 控制器层

1. 员工的处理器类

```
package com.ssmbook2020.controller;

import com.ssmbook2020.entity.Employee;
import com.ssmbook2020.entity.PageBean;
import com.ssmbook2020.service.EmployeeService;
import org.springframework.beans.factory.annotation.Autowired;
import org.springframework.web.bind.annotation.RequestBody;
import org.springframework.web.bind.annotation.RequestMapping;
import org.springframework.web.bind.annotation.RestController;
import java.util.Arrays;
import java.util.Map;

@RestController  //相当于@Controller 和@ResponseBody 两个注解
@RequestMapping("/employee")
public class EmployeeController {

    @Autowired
```

```
    private EmployeeService employeeService;

    /**
     * 查询一页数据
     * @param param 封装查询条件
     */
    @RequestMapping("/findPage")
    public PageBean<Employee> findPage(@RequestBody Map<String, Object>
param) {
        System.out.println("查询条件：" + param);
        return employeeService.findPageBean(param);
    }

    /**
     * 添加员工
     */
    @RequestMapping("/save")
    public int save(@RequestBody Employee employee) {
        System.out.println("添加员工：" + employee);
        return employeeService.save(employee);
    }

    /**
     * 更新员工
     */
    @RequestMapping("/update")
    public int update(@RequestBody Employee employee) {
        System.out.println("更新员工：" + employee);
        return employeeService.update(employee);
    }

    /**
     * 删除员工
     */
    @RequestMapping("/delete")
    public int update(Integer id) {
        System.out.println("删除的员工ID：" + id);
        return employeeService.delete(id);
    }

    /**
     * 删除多个员工
     */
    @RequestMapping("/deleteEmployees")
    public int deleteEmployees(@RequestBody Integer[] employeeIds) {
        System.out.println("删除的员工编号："+Arrays.toString(employeeIds));
        return employeeService.deleteEmployees(employeeIds);
```

```
    }
}
```

2. 部门的处理器类

```
package com.ssmbook2020.controller;

import com.ssmbook2020.entity.Depart;
import com.ssmbook2020.service.DepartService;
import org.springframework.beans.factory.annotation.Autowired;
import org.springframework.stereotype.Controller;
import org.springframework.web.bind.annotation.RequestMapping;
import org.springframework.web.bind.annotation.ResponseBody;

import java.util.List;

@Controller
@RequestMapping("/depart")
public class DepartController {

    @Autowired
    private DepartService departService;

    /**
     * 查询所有部门
     */
    @RequestMapping("/findAll")
    @ResponseBody  //返回格式化好的 JSON 字符串
    public List<Depart> findAllDeparts() {
        return departService.findAllDeparts();
    }
}
```

3. 在浏览器中测试

在 FireFox 浏览器直接访问部门的查询方法 http://localhost:8080/depart/findAll，可以在浏览器上看到打印出来的 JSON 字符串，说明测试成功，如图 11-31 所示。

图 11-31　浏览器上看到的 JSON 数据

至此，服务器后端代码就全部编写完毕了，并且测试通过，接下来开发前端的代码。

11.4 基于 Element 框架的系统开发

11.4.1 什么是 Element

Element 是饿了么公司前端开发团队提供的一套基于 Vue2.0 的网站组件库。它提供了配套设计资源，帮助网站快速成型，使用 Element 前提必须要有 Vue。所谓组件就是组成网页的部件，如超链接、按钮、图片、表格等。Element 的官网地址为 https://element.eleme.cn/#/zh-CN，其首页如图 11-32 所示，读者可以通过这个网站深入学习它每个组件的使用。

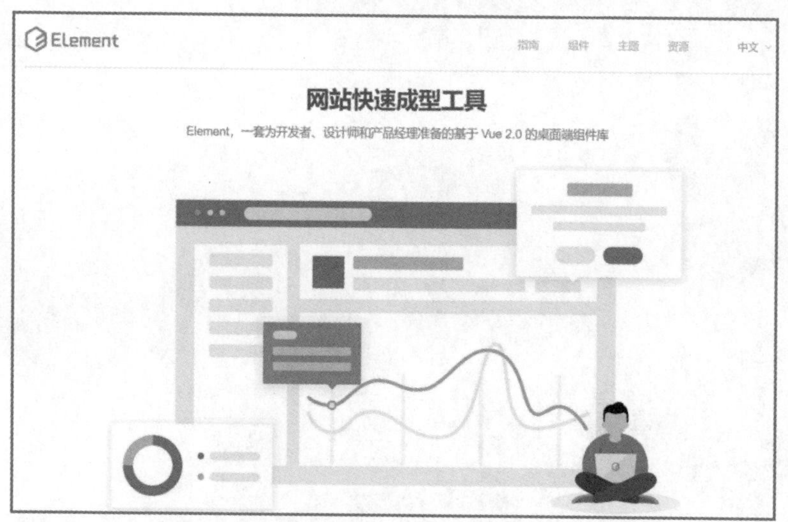

图 11-32　Element 官网首页

11.4.2 Element 快速入门

Element 组件的下载网址为 https://unpkg.com/browse/element-ui@2.14.1/。提供的案例源代码中也有下载好的 Element 组件，可以在 https://github.com/smartxiaomi/ssm2021 中下载。也可以无须下载，直接引用在线组件，后面会讲到该方式。

创建静态 Web 模块的步骤如下。

（1）选择 File 菜单→New 选项→Module 选项。

（2）在 New Module 窗口中选择 Static Web（在 2020 版 IDEA 中，此处为 JavaScript），模块名为 chapter11_elementUI。

（3）将 element-ui 文件夹整个复制到 chapter11_elementUI 目录下，如果是在线使用就跳过这步。

（4）创建 js 目录，将 vue.js 和 axios-0.19.0.js 复制到该目录下。

（5）在根目录下创建 index.html 文件，创建好的结构如图 11-33 所示。

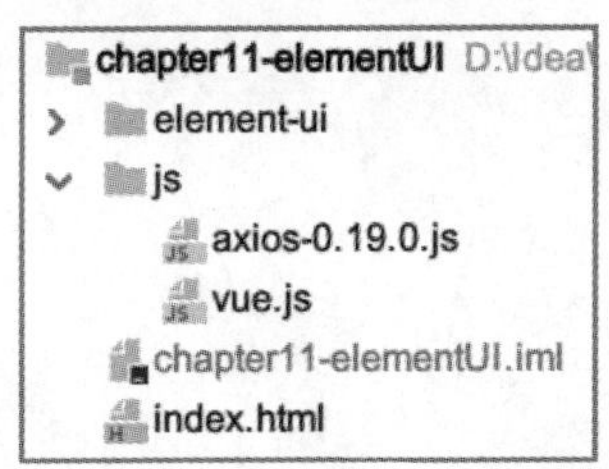

图 11-33　创建 Element 入门模块

（6）在 index.html 中要先导入 Vue 和 Axios，再导入 Element 组件。导入 Element 的两种方式如下。

① 方式一：在线导入。

```
<!-- 引入样式 -->
<link rel="stylesheet" href="https://unpkg.com/element-ui/lib/theme-chalk/index.css">
<!-- 引入组件库 -->
<script src="https://unpkg.com/element-ui/lib/index.js"></script>
```

② 方式二：使用本地下载的组件。

```
<!-- 引入样式 -->
<link rel="stylesheet" href="element-ui/lib/theme-chalk/index.css">
<!-- 引入组件库 -->
<script src="element-ui/lib/index.js"></script>
```

11.4.3　Element 第一个案例

要结合在线文档来学习 Element，因为前端的很多代码只需要直接使用即可，在线文档的地址为 https://element.eleme.cn/#/zh-CN/component/installation。Element 在线文档首页如图 11-34 所示。

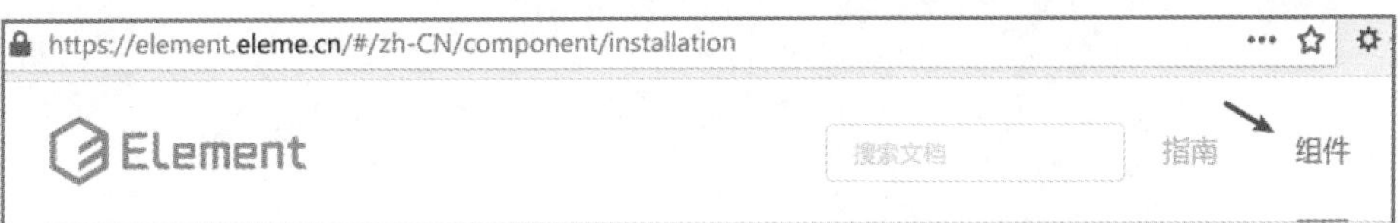

图 11-34　Element 在线文档首页

入门案例中用到了 Container 布局容器和 Button 按钮，其运行效果如图 11-35 所示。

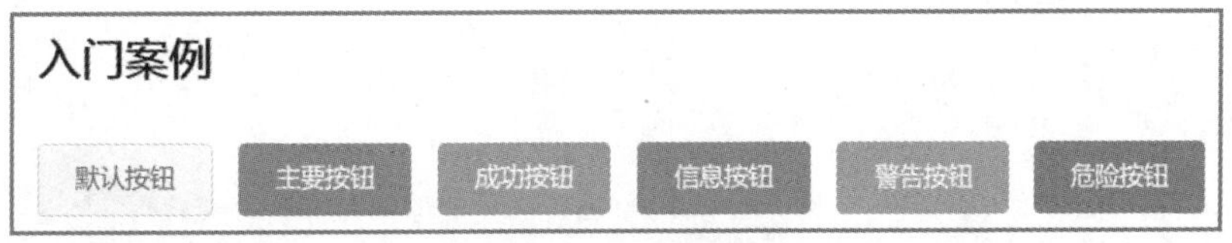

图 11-35　入门案例的效果

实现案例效果的代码如下。

```
<!DOCTYPE html>
<html lang="en">
<head>
    <meta charset="UTF-8">
    <title>ElementUI 入门案例</title>
    <!--导入 vue-->
    <script type="text/javascript" src="js/vue.js"></script>
    <!--导入 axios-->
    <script src="js/axios-0.19.0.js"></script>
    <!-- 引入样式 -->
    <link rel="stylesheet" href="element-ui/lib/theme-chalk/index.css">
    <!-- 引入组件库 -->
    <script src="element-ui/lib/index.js"></script>

</head>
<body>
<!-- 创建 Vue 的容器 -->
<div id="app">
    <!-- 页面容器，由头部和主体两部分组成 -->
    <el-container>
        <!-- 头部 -->
        <el-header><h1>入门案例</h1></el-header>
        <!-- 主体 -->
        <el-main>
            <!--代表一行-->
            <el-row>
                <!-- 不同的按钮样式 -->
                <el-button>默认按钮</el-button>
                <el-button type="primary">主要按钮</el-button>
                <el-button type="success">成功按钮</el-button>
                <el-button type="info">信息按钮</el-button>
                <el-button type="warning">警告按钮</el-button>
                <el-button type="danger">危险按钮</el-button>
            </el-row>
        </el-main>
    </el-container>
</div>

<script type="text/javascript">
    /* 创建 Vue 对象*/
    new Vue({
        el: "#app"
    });
</script>
</body>
</html>
```

在 Element 中提供了大量的组件，如 Table 表格、Progress 进度条、Tree 树形控件和 Pagination 分页等。有些组件的使用是比较复杂的，它不仅是一个 UI 界面的组件，而且提供了不少可直接引用的功能强大的事件处理函数。

11.4.4　使用 Element 实现员工系统的表示层

使用 Element 实现员工系统表示层的步骤如下。

1. 导入 Element 组件库

（1）将 element-ui 和 js 两个目录复制到 webapp 目录下，现在的工程结构如图 11-36 所示。

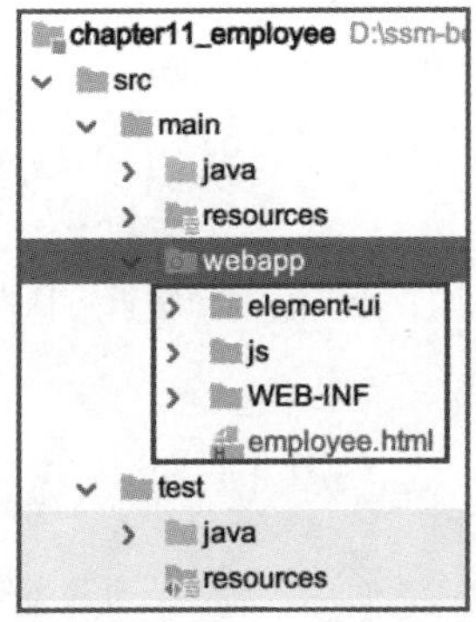

图 11-36　导入 element-ui 和 js 目录后的工程结构

（2）创建 employee.html 文件，整个文件由以下几部分组成。

① 最上面一行是查询条件，如图 11-37 所示。

图 11-37　查询条件

② 中间区域是表格部分，如图 11-38 所示。

	序号	姓名	性别	工资	生日	年龄	所在部门
	1	程伟锋	男	¥3500	1993年11月23日	27	生产部
	2	漆艾林	男	¥5000	1995年02月10日	25	市场部
	3	唐杨斌	男	¥2000	1985年10月04日	35	财务部

图 11-38　表格部分

③ 下面是分页部分，如图 11-39 所示。

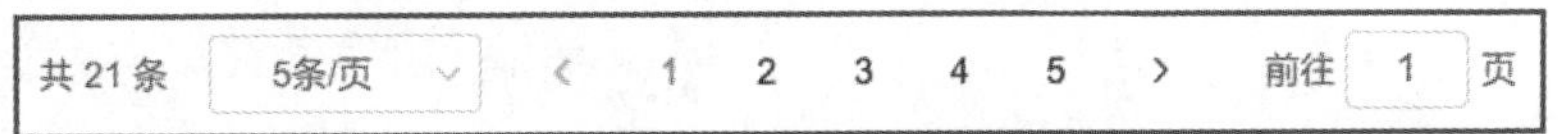

图 11-39　分页部分

④ 最后面是添加和编辑对话框，默认是不显示的，如图 11-40 所示。

图 11-40　添加和编辑对话框

HTML 网页代码如下。

```
<!DOCTYPE html>
<html lang="en">
<head>
    <meta charset="UTF-8">
    <title>显示员工列表</title>
    <!--导入 vue-->
    <script type="text/javascript" src="js/vue.js"></script>
    <!-- 引入样式 -->
    <link rel="stylesheet" href="element-ui/lib/theme-chalk/index.css">
    <!-- 引入组件库 -->
    <script src="element-ui/lib/index.js"></script>
    <!--导入 axios-->
    <script src="js/axios-0.19.0.js"></script>

    <style>
        /* 修改默认的样式 */
        .el-main {
            padding: 10px;
        }

        .el-row {
            margin-bottom: 10px;
        }
    </style>
</head>
<body>
<div id="app">
```

```
        <!-- 一行的查询按钮 -->
        <el-row>
            <!-- 一行分成 24 列 -->
            <el-col :span="24">
                <!-- inline: 以一行的方式显示，model: 绑定的提交参数对象 -->
                <el-form :inline="true" :model="param" size="small">
                    <!-- @change 值发生变化就查询，调用 changeCondition 方法 -->
                    <el-form-item label="姓名">
                        <!-- 绑定 param.name 属性 -->
                        <el-input style="width:150px"
                                  clearable @change="changeCondition" v-model=
"param.name"
                                  prefix-icon="el-icon-search" placeholder="输入查
询姓名"></el-input>
                    </el-form-item>
                    <el-form-item label="性别">
                        <!-- 绑定 param.sex 属性 -->
                        <el-select clearable v-model="param.sex" placeholder="请
选择" style="width:100px" @change="changeCondition">
                            <el-option label="男" value="男"></el-option>
                            <el-option label="女" value="女"></el-option>
                        </el-select>
                    </el-form-item>
                    <el-form-item label="部门">
                        <!--
                        clearable 有清除按钮
                        filterable 可以输入文字过滤
                        multiple 可以多选
                        multiple-limit 最多选 3 个(注意：前面要加冒号，否则报错) -->
                        <el-select clearable filterable multiple
                                   :multiple-limit="3"
                                   v-model="param.departId" placeholder="请选择一个
或多个部门"
                                   style="width: 250px" @change="changeCondition">
                            <!-- 异步加载所有部门并且显示 -->
                            <el-option v-for="depart in departs":key= "depart.id":
 label="depart.name" :value="depart.id"></el-option>
                        </el-select>
                    </el-form-item>
                    <el-form-item label="出生日期">
                        <!-- 绑定 param.birthday 属性 -->
                        <el-date-picker style="width: 280px"
                                        v-model="param.birthday"
                                        type="monthrange" @change="changeCondition"
                                        range-separator="至"
                                        start-placeholder="开始月份"
                                        end-placeholder="结束月份"
```

```
                                            format="yyyy-mm-dd"
                                            value-format="yyyy-mm-dd">
                              </el-date-picker>
                          </el-form-item>
                          <!-- 三个按钮 -->
                          <el-form-item>
                              <el-button type="success" plain size="small" icon="el-
icon-document-delete" @click="clearCondition">清除</el-button>
                              <el-button type="danger" plain size="small" icon="el-
icon-delete" @click="deleteEmployees">删除</el-button>
                              <el-button type="primary" plain size="small" icon="el-
icon-document" @click="showAddDialog">添加</el-button>
                          </el-form-item>
                      </el-form>
                  </el-col>
              </el-row>

              <!--表格显示数据部分-->
              <el-row>
                  <el-col :span="24">

                      <!--
                      @selection-change 前面的复选框发生变化激活的事件,
                      data 属性用于绑定 pageBean 对象中的 data 属性, pageBean 异步加载获取
                      -->
                      <el-table :data="pageBean.data" max-height="95%" style="width:
100%" @selection-change="handleSelectionChange"
                              :header-cell-style="{background:'#02b5fc',color:'white'}"
border>

                          <!--表格前面的复选框-->
                          <el-table-column type="selection" width="55"></el-table-
column>

                          <!-- index 是生成序号的方法 -->
                          <el-table-column
                                  label="序号" type="index" width="60" :index="indexMethod">
                          </el-table-column>

                          <!--每一列绑定一行对象中相应的属性 -->
                          <el-table-column
                                  prop="name"
                                  label="姓名">
                          </el-table-column>

                          <el-table-column
                                  prop="sex"
```

```
                    label="性别">
                </el-table-column>

                <el-table-column
                    prop="salary"
                    label="工资" :formatter="formatSalary">
                </el-table-column>

                <el-table-column
                    prop="birthday" label="生日">
                </el-table-column>

                <el-table-column
                    prop="age" label="年龄">
                </el-table-column>

                <el-table-column
                    prop="depart.name" label="所在部门">
                </el-table-column>

                <el-table-column label="操作" width="200">
                  <!-- scope 表示表格中所有的数据对象 -->
                  <template slot-scope="scope">
                    <!-- scope.row 表示这一行数据对象 -->
                    <el-button
                        size="mini" icon="el-icon-edit" type="success"
plain
                        @click="handleEdit(scope.$index, scope.row)">
编辑
                    </el-button>
                    <el-button
                        size="mini"
                        type="warning" icon="el-icon-delete" plain
                        @click="handleDelete(scope.$index, scope.row)">
删除
                    </el-button>
                  </template>
                </el-table-column>

            </el-table>

        </el-col>
    </el-row>

    <!--分页部分-->
    <el-row>
        <el-col :span="24">
```

```
            <!--
            @size-change 表示修改页面大小激活的函数
            @current-change 表示修改当前页激活的函数
            current-page 表示当前页号
            page-sizes 表示页面大小的选项值
            page-size 表示默认每页的大小
            layout 表示分页组件的组成元素
            total 表示总页数
            -->
            <el-pagination
                  @size-change="handleSizeChange"
                  @current-change="handleCurrentChange"
                  :current-page="1"
                  :page-sizes="[3,5,10]"
                  :page-size="5"
                  layout="total, sizes, prev, pager, next, jumper"
                  :total="pageBean.count">
            </el-pagination>
         </el-col>
      </el-row>

      <!--
      新增和修改对话框
      close-on-click-modal:是否可以通过单击其他位置关闭对话框
       -->
      <el-dialog title="员工信息" :visible.sync="dialogFormVisible" width=
"30%" @close="employeeCancel"
               :close-on-click-modal="false">
         <!-- status-icon 为输入框添加了表示校验结果的反馈图标-->
         <el-form :model="form" :rules="rules" label-width="100px" size=
"small" ref="employeeForm" status-icon>
            <!-- 绑定 form 对象中的各个属性 -->
            <el-form-item label="姓名" prop="name">
               <el-input v-model="form.name" autocomplete="off" placeholder=
"请输入姓名" prefix-icon="el-icon-user" style="width:85%"></el-input>
            </el-form-item>

            <el-form-item label="性别" prop="sex">
               <el-radio v-model="form.sex" label="男">男</el-radio>
               <el-radio v-model="form.sex" label="女">女</el-radio>
            </el-form-item>

            <el-form-item label="工资" prop="salary">
               <el-input v-model="form.salary" type="text" autocomplete=
"off" placeholder="请输入工资" style="width:85%">
                  <template slot="prepend" style="width:30px;">¥</template>
               </el-input>
```

```
            </el-form-item>

            <el-form-item label="生日" prop="birthday">
                <!--要指定值为 yyyy 年 mm 月 dd 日格式，Spring 会自动封装到
java.sql.Date 类型中-->
                <el-date-picker style="width:85%"
                                v-model="form.birthday"
                                type="date" format="yyyy-mm-dd" value-format=
"yyyy 年 mm 月 dd 日"
                                placeholder="选择或输入日期 (格式：年-月-日)">
                </el-date-picker>
            </el-form-item>

            <!--部门从部门表中异步加载到下拉列表中-->
            <el-form-item label="部门" prop="departId">
                <el-select v-model="form.departId" placeholder="选择或输入部门"
filterable style="width:85%">
                    <!-- 值是 id，显示的文本是 name -->
                    <el-option
                            v-for="depart in departs"
                            :key="depart.id"
                            :label="depart.name"
                            :value="depart.id">
                    </el-option>
                </el-select>
            </el-form-item>

        </el-form>
        <div slot="footer" class="dialog-footer">
            <!-- 单击激活相应的方法 -->
            <el-button @click="employeeCancel" icon="el-icon-ice-cream-
round"> 取 消</el-button>
            <el-button type="primary" @click="saveEmployee" icon="el-icon-
milk-tea"> 确 定</el-button>
        </div>
    </el-dialog>
</div>

<!--  导入自己写的 JS 代码-->
<script src="js/employee.js"></script>
</body>
</html>
```

2. 前端 JS 代码

自己写的 JS 代码放在 js/employee.js 文件中，与 HTML 进行了分离，如图 11-41 所示。

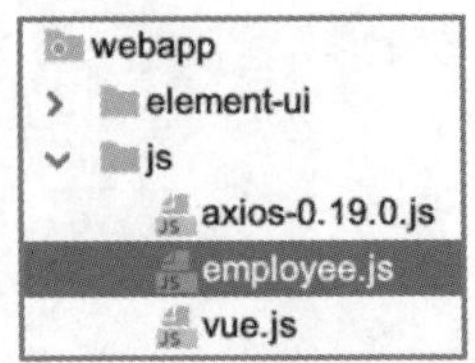

图 11-41　JS 代码位置

JS 代码如下。

```
let app = new Vue({
    el: "#app",
    data() {
        //自定义表单验证规则，验证工资项
        let checkSalary = (rule, value, callback) => {
            if (isNaN(value)) {
                return callback(new Error('请输入数字值'));
            }
            if (value < 0) {
                return callback(new Error('必须大于 0'));
            }
            //验证通过
            return callback();
        };
        return {
            //封装整个页面数据
            pageBean: {},
            //封装查询条件
            param: {
                name: "",
                sex: "",        //查询条件
                departId: [], //部门主键，可以查询多个部门
                birthday: [], //生日范围。注意：这里是一个数组，在服务器端是一个
                              //ArrayList 集合
                current: 1,   //当前页
                size: 5       //每页大小
            },
            //所有的部门集合
            departs: [],
            //前面的复选框
            employeeIds: [],
            //对话框默认不可见
            dialogFormVisible: false,
            //表单项的内容
            form: {
                name: '',
                sex: '男',
                salary: '',
```

```
            birthday: '',
            departId: ''
        },
        //校验规则
        rules: {
            //员工姓名
            name: [
                {required: true, message: '员工姓名不能为空', trigger: 'blur'},
                {min: 2, max: 10, message: '长度在 2 到 10 个字符', trigger: 'blur'}
            ],
            sex: [
                {required: true, message: '请选择性别', trigger: 'blur'}
            ],
            salary: [
                {required: true, message: '请输入工资', trigger: 'blur'},
                //使用自定义规则
                {validator: checkSalary, trigger: 'blur'}
            ],
            birthday: [
                {required: true, message: '请选择日期', trigger: 'blur'}
            ],
            departId: [
                {required: true, message: '选择一个部门', trigger: 'blur'}
            ]
        }
    }
},
methods: {
    //改变页大小
    handleSizeChange(val) {
        this.param.size = val;
        //查询所有页面
        this.findAllEmployees();
    },
    //改变当前页
    handleCurrentChange(val) {
        this.param.current = val;
        //查询所有页面
        this.findAllEmployees();
    },
    //编辑按钮：索引号，这一行的数据对象
    handleEdit(index, row) {
        //打开编辑窗口
        this.dialogFormVisible = true;
        //显示数据在窗口中，直接赋值会导致表格中显示的数据也发生变化
        Object.assign(this.form, row);   //复制属性
    },
```

```
//删除按钮
handleDelete(index, row) {
    this.$confirm('确定要删除员工 ' + row.name + ' 吗?', '删除操作', {
        type: 'warning'
    }).then(() => {
        //后台访问服务器删除
        axios.get("employee/delete?id=" + row.id).then((resp) => {
            //在控制台输出结果
            console.log(resp.data);
            if (resp.data > 0) {
                this.$message.success("成功删除员工" + row.name);
                //重新加载员工
                this.findAllEmployees();
            } else {
                this.$message.error("删除员工" + row.name + "失败");
            }
        });
    }).catch(() => {
    });
},
//前面的复选框发生变化
handleSelectionChange(val) {
    //先清空
    this.employeeIds = [];
    //获取的是整个行的数组
    for (let row of val) {
        //只得到每行的 id，添加到集合中
        this.employeeIds.push(row.id);
    }
},
//显示编辑的窗口
showAddDialog() {
    //设置为可见
    this.dialogFormVisible = true;
    //将 form 清空
    this.form = {
        name: '',
        sex: '男',
        salary: '',
        birthday: '',
        departId: ''
    }
},
//添加员工的取消按钮
employeeCancel() {
    this.dialogFormVisible = false;
    //表单内容重置，清除校验规则
```

```
            this.$refs.employeeForm.resetFields();
        },
        //添加员工的确定按钮
        saveEmployee() {
            //提交前再次验证，$refs 引用表单上 ref 属性中名字，调用 validate 方法验证
            //整个表单
            this.$refs["employeeForm"].validate((valid) => {
                //验证通过
                if (valid) {
                    //如果有 id 值则是更新，没有就是添加
                    console.log(this.form);
                    //更新
                    if (this.form.id) {
                        axios.post("employee/update",this.form).then((resp)=>{
                            let row = resp.data;
                            //成功以后才查询员工
                            if (row > 0) {
                                this.$message.success('成功更新员工'+
this.form.name + ' 的信息');
                                //关闭信息框
                                this.dialogFormVisible = false;
                                //表单内容重置，清除校验规则
                                this.$refs.employeeForm.resetFields();
                                //重新加载所有员工
                                this.findAllEmployees();
                            } else {
                                this.$message.error('更新员工 ' + this.form.name
+ ' 失败');
                            }
                        });
                    }
                    //添加
                    else {
                        axios.post("employee/save", this.form).then((resp) => {
                            let row = resp.data;
                            //添加成功
                            if (row > 0) {
                                this.$message.success('成功添加员工 ' +
this.form.name + ' 的信息');
                                //关闭信息框
                                this.dialogFormVisible = false;
                                //表单内容重置，清除校验规则
                                this.$refs.employeeForm.resetFields();
                                //重新加载所有员工
                                this.findAllEmployees();
                            } else {
                                this.$message.error('添加员工 ' + this.form.name
```

```
+ ' 失败');
                            }
                        });
                    }
                }
            });

        },
        //第一列的行号
        indexMethod(index) {
            return (this.param.current - 1) * this.param.size + 1 + index;
        },
        //格式化工资列
        formatSalary(row, column, cellValue, index) {
            return "¥" + cellValue;
        },
        //清除查询条件按钮
        clearCondition() {
            this.param.name = "";
            this.param.sex = "";
            this.param.departId = [];
            this.param.birthday = [];
            //无条件查询
            this.findAllEmployees();
        },
        //所有的查询条件发生变化调用这个方法，因为绑定了 condition 中的属性
        changeCondition() {
            //先跳到第一页，以免出现显示不了的 bug
            this.param.current = 1;
            //再查询
            this.findAllEmployees();
        },
        //分页查询所有联系人
        findAllEmployees() {
            axios.post("employee/findPage", this.param).then(resp => {
                //获取 PageBean 对象
                this.pageBean = resp.data;
            });
        },
        //查询所有的部门
        findAllDeparts() {
            axios.post("depart/findAll").then(resp => {
                //得到所有的部门集合
                this.departs = resp.data;
            });
        },
        //删除多个员工
```

```
        deleteEmployees() {
            let num = this.employeeIds.length;
            //判断是否有选中
            if (num == 0) {
                this.$alert('请至少选中一个要删除的员工', '批量删除', {
                    type: 'info'
                });
            } else {
                this.$confirm("确定要删除这" + num + "个员工吗?", "批量删除", {
                    type: 'warning'
                }).then(() => { //单击确定按钮
                   axios.post("employee/deleteEmployees",this.employeeIds).
then((resp) => {
                        let row = resp.data;
                        if (row > 0) {
                            this.$message.success("成功删除" + row + "个员工");
                            //重新加载
                            this.findAllEmployees();
                        } else {
                            this.$message.error("删除员工失败");
                        }
                    });
                });
            }
        }
    },
    //Vue 对象创建完毕以后自动调用的方法
    created() {
        //查询所有员工
        this.findAllEmployees();
        //查询所有的部门，给下拉列表用
        this.findAllDeparts();
    }
});
```

3. 测试运行

因为在 webapp 目录下删除了 index.jsp 文件，为了让项目启动就需要运行 employee.html，可以在 web.xml 中添加以下欢迎页面的配置。

```
<!-- 添加欢迎页面 -->
<welcome-file-list>
    <welcome-file>employee.html</welcome-file>
</welcome-file-list>
```

如果在控制台测试了乱码的问题，可以再添加-Dfile.encoding=GBK 参数，这是因为 Maven 中 Tomcat7 的插件默认使用的是 GBK 编码，如图 11-42 所示。

运行成功以后，在控制台会看到 SQL 语句和查询结果。

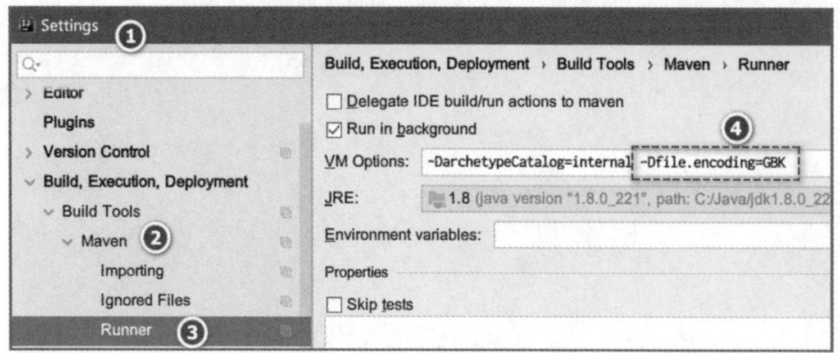

图 11-42　控制台测试乱码问题

```
查询条件：{name=, sex=, departId=[], birthday=[], current=1, size=5}
Creating a new SqlSession
SqlSession [org.apache.ibatis.session.defaults.DefaultSqlSession@46ee52fd] was not registered for synchronization because synchronization is not active
JDBC Connection [com.mysql.jdbc.JDBC4Connection@66130f25] will not be managed by Spring
==>  Preparing: SELECT e.*, TIMESTAMPDIFF(YEAR,birthday,NOW()) AS age, d.id did, d.name dname FROM employee e LEFT JOIN depart d ON e.depart_id = d.id ORDER BY e.id LIMIT ?,?
==> Parameters: 0(Integer), 5(Integer)
<==    Columns: id, name, sex, salary, birthday, depart_id, age, did, dname
<==        Row: 1, 程伟锋, 男,  3500, 1993-11-23, 2, 27, 2, 生产部
<==        Row: 2, 漆艾林, 男,  5000, 1995-02-10, 5, 25, 5, 市场部
<==        Row: 3, 唐杨斌, 男,  2000, 1985-10-04, 4, 35, 4, 财务部
<==        Row: 5, 陈香玉, 女,  6888, 1999-09-04, 7, 21, 7, 研发部
<==        Row: 6, 邹丽娜, 女, 12000, 1982-02-25, 7, 38, 7, 研发部
<==      Total: 5
Closing non transactional SqlSession 小结和习题
```

11.5　本章小结

本章通过综合应用案例贯穿了前面学习的 SSM 知识点，同时也增加了前端框架知识。

本章首先介绍了 Maven 的基本使用，然后搭建了 SSM 的开发环境。前端使用 ElementUI 框架来实现，它是基于 Vue 的一个前端框架。使用前端框架可以起到事半功倍的效果，制作出来的页面既美观，又减少了工作量。通过本章案例，将前面所学的 SSM 的知识点进行了综合应用。

习题 11

将本章综合案例自行开发并正确部署运行。

参 考 文 献

[1] WALLS C. Spring 实战[M]. 6 版. 张卫滨，吴国浩，译. 北京：人民邮电出版社，2022.

[2] CARNELL J. Spring 微服务实战[M]. 2 版. 陈文辉，译. 北京：人民邮电出版社，2022.

[3] 彭之军，刘波. Java EE SSH 框架应用开发项目教程[M]. 2 版. 北京：电子工业出版社，2019.

[4] 陈恒，李正光. SSM 开发实战教程[M]. 北京：清华大学出版社，2022.

[5] 彭之军，刘波. Java EE Spring MVC 与 MyBatis 企业开发实战[M]. 北京：电子工业出版社，2019.

[6] 陈雄华. 精通 Spring 4.x 企业应用开发实战[M]. 北京：电子工业出版社，2017.

[7] DEINUM M，RUBIOD，LONG J. Spring 5 攻略[M]. 张龙，译. 北京：人民邮电出版社，2021.

[8] 陈恒，李正光. SSM+Spring Boot+Vue.js 3 全栈开发从入门到实战[M]. 北京：清华大学出版社，2022.

[9] 林信良. JSP & Servlet 学习笔记[M]. 3 版. 北京：清华大学出版社，2019.

图书资源支持

感谢您一直以来对清华版图书的支持和爱护。为了配合本书的使用，本书提供配套的资源，有需求的读者请扫描下方的“书圈”微信公众号二维码，在图书专区下载，也可以拨打电话或发送电子邮件咨询。

如果您在使用本书的过程中遇到了什么问题，或者有相关图书出版计划，也请您发邮件告诉我们，以便我们更好地为您服务。

我们的联系方式：

地　　址：北京市海淀区双清路学研大厦 A 座 714

邮　　编：100084

电　　话：010-83470236　010-83470237

客服邮箱：2301891038@qq.com

QQ：2301891038（请写明您的单位和姓名）

资源下载：关注公众号“书圈”下载配套资源。

资源下载、样书申请

书圈

图书案例

清华计算机学堂

观看课程直播